Bilingual Course of Chemical Automation and Instrumentation

化工自动化及仪表双语教程

张 泉 主编 王 莉 副主编

化学工业出版社

·北京·

内容提要

本教材根据高职高专化工专业和自动化类专业的教学要求，以化工自动化及仪表方面的基础知识为教学主轴，主要内容包括化工过程变量的检测方法和检测仪表、控制器与控制阀、简单控制系统与复杂控制方案三大部分。

通过本教材的学习，学生应能够进行相关英语技术资料的查阅和搜索。本教材主要面向高职高专自动化专业及化工专业学生和企业技术人员，旨在为其后续的英文信息收集、资料撰写等工作打下基础。

图书在版编目（CIP）数据

化工自动化及仪表双语教程/张泉主编. —北京：化学工业出版社，2020.8

ISBN 978-7-122-36726-6

Ⅰ.①化… Ⅱ.①张… Ⅲ.①化工过程-自动控制系统-双语教学-高等职业教育-教材②化工仪表-双语教学-高等职业教育-教材 Ⅳ.①TQ056

中国版本图书馆 CIP 数据核字（2020）第 082218 号

责任编辑：刘　哲　葛瑞祎　　装帧设计：刘丽华
责任校对：边　涛

出版发行：化学工业出版社（北京市东城区青年湖南街 13 号　邮政编码 100011）
印　　装：三河市延风印装有限公司
787mm×1092mm　1/16　印张 8　字数 192 千字　2020 年 9 月北京第 1 版第 1 次印刷

购书咨询：010-64518888　　售后服务：010-64518899
网　　址：http://www.cip.com.cn
凡购买本书，如有缺损质量问题，本社销售中心负责调换。

定　　价：32.00 元

Preface 前言

自动化及仪表技术的应用和发展，是推动化工生产的强大动力。过程自动化领域对专业人才的要求已经不仅仅限于掌握一定的技能，专业精通、英语能力强、具有国际视野、有终身学习潜力的复合型人才更受欢迎，具有外语技术资料的阅读和搜索能力成为一种必需。本书的编写宗旨是，以化工自动化及仪表的基础知识为主轴，通过合理的章节安排，对化工生产过程中涉及的常用仪表和基本控制方案，以中英文对照的方式进行介绍。本书条理清晰、难易适中、篇幅适当、语言规范，没有繁琐的公式推导和理论分析，重要词汇的复现率高，读者易入门并能较好地激发学习兴趣。

本书主要面向高职化工、自动化类学生和企业生产技术人员，旨在帮助读者扩展对本学科基本关键技术的认识和实际应用能力，扩充化工自动化及仪表方面的英语词汇量，提高技术资料的阅读和翻译水平，为后续的英文信息收集、技术研发、文件撰写等工作打下基础。本书也可作为具有一定英语基础的机电、电气等相关技术人员的参考书。

本书共分 9 章，内容范围包括化工过程变量的检测方法和检测仪表、控制器与控制阀、简单控制系统与复杂控制方案三大部分，参考学时为 35 学时。每一章均按内容不同进行分节，每一节的英文教学内容都附有对应的中文翻译。在每一章的最后配有该章节的重要生词和短语，方便读者快速查找。

本书由南京科技职业学院张泉主编并统稿，王莉为副主编。第 2、3、4、5、7 章由张泉编写，第 1、6 章由王莉编写，第 8、9 章由朱玉奇编写。本书的出版，得到了化学工业出版社的大力支持。同时，河海大学能源与电气学院卢新彪副教授和南京科远自动化集团股份有限公司吕德宏高级工程师参与了本书的内容制订并给予编写指导，南京科技职业学院王永红教授也提出了很多宝贵意见和建议，在此向他们表示衷心感谢！本书的出版，得到了江苏省高等职业教育高水平骨干专业建设项目的经费支持。

编者水平所限，不妥之处请读者予以指正。

编者

2020 年 4 月

Contents 目录

Chapter 1

Process Measurement Fundamentals

第 1 章

过程测量基本原理

1.1 Introduction

Measurement and control systems monitor and control processes that otherwise would be difficult to operate efficiently and safely while meeting the requirements for high quality and low cost. All process control starts with measurement, and the quality of control obtained can never be better than the quality of the measurement on which it is based.

Measurement refers to the conversion of the process variable (PV) into an analog or digital signal that can be used by the control system. The device that performs the initial measurement is called a sensor or instrument. Typical measurements are pressure, level, temperature, flow, position, and speed. The result of any measurement is the conversion of a dynamic variable into some proportional information that is required by the other devices in the process control loop.

Because sensors are the first device in the control loop to measure the process variable, they are also called primary elements. Primary elements are devices that sense the process variable and translate that sensed quantity into an analog representation (electrical voltage, current, resistance; mechanical force, motion, etc.). Examples: thermocouple, thermistor, bourdon tube, pressure sensing diaphragms, orifice plate, RTDs. For example, With an RTD, as the temperature of a process fluid surrounding the RTD rises or falls, the electrical resistance of the RTD increases or decreases a proportional amount. The resistance is measured, and from this measurement, temperature is determined.

A transducer is a device that receives information in the form of one or more physical quantities, modifies the information or its form if required, and produces a resultant output signal. The output signal of the transducer is not a standard 4 to 20mA signal, but a dimensional signal. For example, inside a capacitance pressure device, a transducer converts changes in pressure into a proportional change in capacitance.

A converter is a device that converts one type of signal into another type of signal. For example, a converter may convert current into voltage or an analog signal into a digital signal.

A transmitter is a device that converts a reading from a sensor or transducer into a standard signal and transmits that signal to a monitor or controller. In the chemical process, typical transmitters include flow transmitters, pressure transmitters, level transmitters, and temperature transmitters.

【译文】

1.1 简介

测量与控制系统监视和控制着过程，否则生产将难以保证高效安全地运行，同时也无法满足高质量和低成本的要求。有效的过程控制都是基于正确的测量，测量质量的高低将决定控制的优劣。

测量是指将过程变量转换为控制系统可以使用的模拟或数字信号。执行测量的设备称为传感器或仪表。典型的测量值有压力、液位、温度、流量、位置和速度等。任何测量的结果都是将动态变量转换为过程控制回路中其他装置所需的一些对应信息。

由于传感元件是控制回路中测量过程变量的首个装置，因此它们也被称为一次元件。一次元件是测量过程变量，并将该测量值转换为模拟示值（电压、电流、电阻、机械力、运动等）的设备，如热电偶、热敏电阻、弹簧管、压力感应膜片、孔板、热电阻等。例如，对于热电阻，随着它周围的过程流体温度的升高或降低，热电阻的阻值会成比例地增加或减少，通过测量电阻值就可以确定温度。

传感器是一种接收一个或多个物理量形式的信息，转换成所需的形式并产生输出信号的设备。传感器的输出信号不是标准的 4～20mA 信号，而是带量纲的信号。例如，在电容压力装置中，传感器将压力值转换为对应的电容变化值。

转换器是一种将某类信号转换为另一类信号的设备。例如，转换器可以将电流转换成电压，或将模拟信号转换成数字信号。

变送器是将传感元件或传感器的读数转换为标准信号，并将该信号传输到监视器或控制器的一种设备。在化工过程中，典型的变送器包括流量变送器、压力变送器、液位变送器和温度变送器等。

1.2 Basic Measurement Performance Terms

There are a number of criteria that must be satisfied when specifying process measurement equipment. All instruments, irrespective of their measurement requirements, exhibit the same characteristics such as accuracy, span, etc. This section explains and demonstrates an interpretation of basic measurement performance terms.

Accuracy The accuracy specified by a device is the amount of error that may occur when measurements are taken. It determines how precise or correct the measurements are to the actual value and is used to determine the suitability of the measuring equipment. Accuracy can be expressed as the error in units of the measured value, such as −2kPa; or accuracy can be expressed in percent of span, such as ±0.5% of span.

Sensitivity The ratio of the change in output magnitude to the change of the input that causes it after the steady-state has been reached. It is expressed as a ratio with the units of measurement of the two quantities stated. The greater the output signal change from the transducer's transmitter for a given input change, the greater the sensitivity of the measuring element.

Resolution Precision is related to resolution, which is defined as the smallest change of input that results in a significant change in transducer output.

Repeatability The closeness of agreement between a number of consecutive measurements of the output for the same value of input under identical operating conditions, approaching from the same direction for full range transverses is usually expressed as repeatability in percent of span. Continuous control applications can be affected by variations due to repeatability. When a control system sees a change in the parameter it is controlling, it will adjust its output accordingly. However if the change is due to the repeatability of the measuring device, then the controller will over-control.

Linearity Linearity expresses the deviation of the actual reading from a straight line. If all outputs are in the same proportion to corresponding inputs over a span of values, then input-output plot is a straight line else it will be non linear (see Figure 1-1).

Hysteresis The maximum deviation between the two curves obtained from the upward and downward measurement values in the entire range of the meter. its output varies as shown in Figure 1-2.

Range Any instrument will be able to measure values of a process variable only between fixed lower and upper limits. Range is the region between stated upper and lower range values of which the quantity is measured. Example: a differential pressure range is stated as 10～200kPa.

Span Span should not be confused with range, although the same points of reference are used. Span is the Algebraic difference between the upper and lower range values, it is usually a positive number. Example: If the range is stated as 10～200kPa then the span is 200－10＝190kPa.

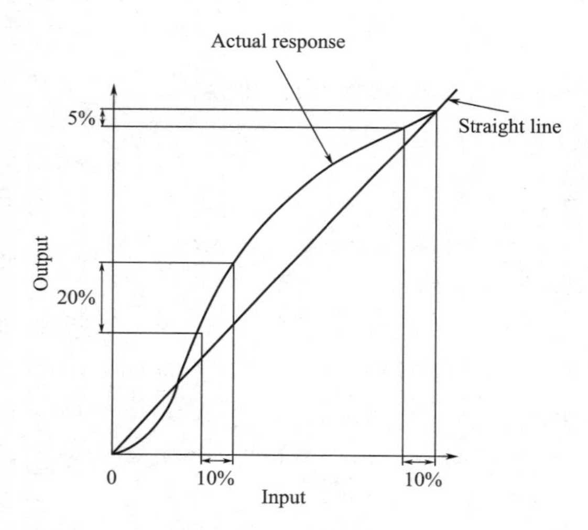

Figure 1-1 Linearity

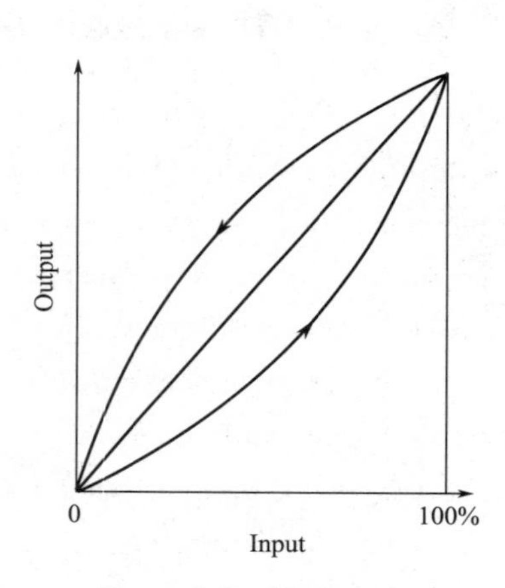

Figure 1-2 Hysteresis

In addition to these criteria mentioned above, there are many other factors that must be considered before selecting a specific method for measuring a process variable for a specific process:

- The normal range over which the PV may vary
- The sensor dynamics required
- The reliability that is required
- The costs involved, including installation and operating costs as well as purchase costs
- The installation requirements and problems, such as size and shape restraints, remote transmission, corrosive fluids, explosive mixtures, etc.

【译文】

1.2 基本的测量性能指标

在选择过程测量设备时，必须满足许多标准。所有仪表，无论其测量要求如何，都具有相同的一些特性指标，例如精度、量程等。本节介绍这些基本的测量性能指标。

精度 测量设备的精度是进行测量时可能发生的误差量。它确定了测量值对实际值的精密度或准确度，并用于确定该测量设备的适用性。精度可以表示为以测量值为单位的误差，例如－2kPa；也可以用量程的百分比表示，例如量程的±0.5%。

灵敏度 仪表在达到稳定状态后输出量（测量值）变化与输入量（被测量）变化的比值称为灵敏度。对于给定的输入变化，传感器的输出信号变化越大，灵敏度就越大。

分辨率 分辨率与精密度有关，分辨率被定义为导致传感器输出发生显著变化的最小输入变化。

重复性 重复性是在相同的操作条件下，对于相同范围的输入，在同一方向上做连续多次变化时，输出的多个结果之间的一致性接近程度。通常以量程的百分比表示重复性。连续控制过程可能会受到测量仪表重复性变化的影响。当控制系统发现被控变量发生变化时，控制器将相应地调整其输出，但是如果被控变量的变化是由于测量仪表的可重复性引起的，那么控制器将会出现过度控制。

线性度 线性度表示实际读数与直线的偏差。如果在一段量程内，所有输出与相应输入的比例相同，则输入与输出关系为直线，否则为非线性（图 1-1）。

回差 回差是在仪表的整个量程中，输入由小到大和由大到小变化得到的两条输出曲线之间的最大偏差。其输出如图 1-2 所示。

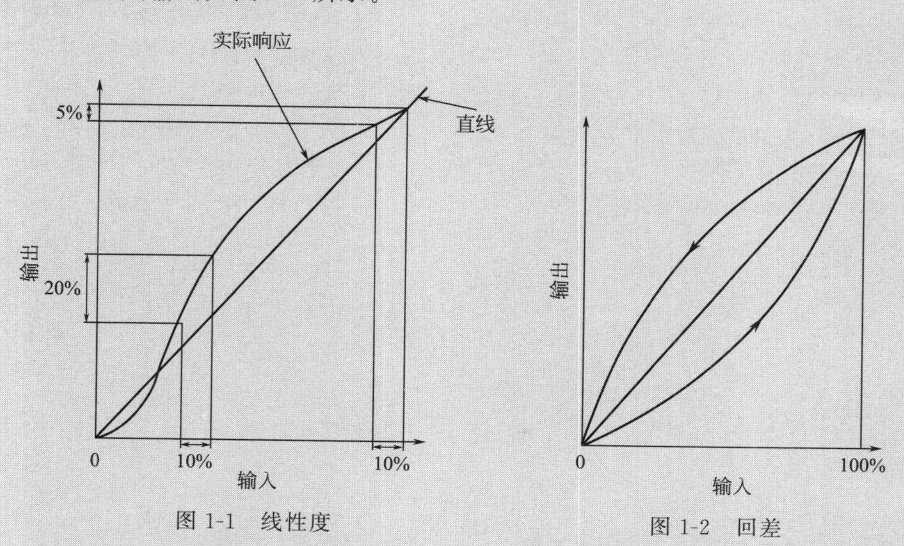

图 1-1 线性度

图 1-2 回差

测量范围 任何检测仪表只能测量固定的下限值和上限值之间的过程变量的值。测量范围是测量值所规定的上限值和下限值之间的区域。例如，差压范围为 10～200kPa。

量程 量程不应与测量范围混淆，尽管使用了相同的参考点。量程是上、下取值范围的代数差，通常是一个正数。例如：如果测量范围表示为 10～200kPa，则量程为 200－10＝190kPa。

除了以上提到的这些标准外，在为特定的过程选择过程变量的测量方法时，还必须考虑许多其他因素：

- 过程变量变化的正常范围
- 传感器的动态特性
- 可靠性
- 所涉及的成本，包括安装、操作维护成本及购买成本
- 安装要求和问题，如尺寸和形状的限制、远程传输、腐蚀性液体、爆炸性混合物等。

Words and Expressions 词汇和短语

1. measurement /'meʒərmənt/ n. 测量
2. monitor /'mɑːnɪtə/ vt. 监控
3. process variable 过程变量
4. analog /'ænəlɔg/ n. 模拟量
5. digital /'dɪdʒɪtl/ n. 数字量
6. pressure /'preʃə/ n. 压力
7. level /'levl/ n. 液位
8. temperature /'temprətʃə(r)/ n. 温度
9. flow /floʊ/ n. 流量
10. sensor /'sensər/ n. 传感器
11. instrument /'ɪnstrəmənt/ n. 仪表
12. primary element 一次元件
13. thermocouple /'θɜməkʌpl/ n. 热电偶
14. thermistor /θɜ'mɪstə/ n. 热敏电阻
15. bourdon tube 弹簧管
16. diaphragm /'daɪəfræm/ n. 膜片
17. orifice plate 孔板
18. transducer /træns'duːsə/ n. 传感器
19. converter /kən'vɜːrtə/ n. 转换器
20. transmitter /træns'mɪtə/ n. 变送器
21. performance /pər'fɔːrməns/ n. 性能
22. criteria /kraɪ'tɪrɪə/ n. 标准
23. accuracy /'ækjərəsi/ n. 精度
24. sensitivity /ˌsensə'tɪvəti/ n. 灵敏度
25. resolution /ˌrezə'luːʃn/ n. 分辨率
26. repeatability /ri'piːtə'biliti/ n. 重复性
27. linearity /ˌlɪni'ærəti/ n. 线性度
28. hysteresis /ˌhɪstə'risɪs/ n. 回差
29. range /reɪndʒ/ n. 测量范围
30. span /spæn/ n. 量程
31. steady-state 稳态
32. precision /prɪ'sɪʒn/ n. 精密度
33. installation /ˌɪnstə'leɪʃn/ n. 安装

Chapter 2

Temperature Measurement

第2章

温度测量

2.1 Introduction

Temperature is a more easily detected quantity than thermal energy, when we need to measure thermal energy, we do so by measuring temperature and then inferring the desired variable based on the laws of thermodynamics. Temperature is the principal process variable of serious concern to the process industries, chemical, petroleum, petrochemical, polymer, plastic, and large segments of metallurgical and food processors are examples. Temperature control is critical to such processes and operations as chemical reactions and in materials separations, such as distillation, drying, evaporation, absorbing, crystallizing, baking, and extruding. Temperature control also plays a critical role in the safe operation of such facilities.

The most commonly used units of temperature measurement are the Fahrenheit scale and the Celsius scale. The Fahrenheit scale was invented by Daniel G. Fahrenheit and published in 1724. It is still extensively used in the United States, although some industries are gradually converting to Celsius. The Celsius scale was developed by Anders Celsius, a Swedish scientist, in 1742, and it is most widely used.

Degrees fahrenheit (°F), degrees Celsius (℃), and Kelvin (K, used mainly for scientific work) are recognized internationally as scales for measuring temperature. It has been experimentally determined that the lowest possible temperature is −273.15℃. The Kelvin temperature scale was chosen so that its zero is at −273.15℃, and the size of one Kelvin unit was the same as the Celsius degree. Conversion from one scale into the other follows these equations:

$$℉=\left(℃\times\frac{9}{5}\right)+32 \tag{2-1}$$

$$℃=K-273.15 \tag{2-2}$$

Temperature is the most common process variable measured in process control. Due to the vast temperature range that needs to be measured (from absolute zero to thousands of degrees) with spans of just a few degrees and sensitivities down to fractions of a degree, there is a vast range of devices that can be used for temperature measurements.

Physical properties that change with temperature are used to measure temperature. For example, the property of material expansion when heated is used in liquid-in-glass, bimetallics, and filled-system measurement. Early thermometers depended on volumetric changes of gases and liquids with temperature change, and this principle still is exploited, as encountered in industrial gas and liquid filled thermal systems and in the liquid-column thermometer. Although these instruments were accepted widely for many years, the filled-system thermometer has been significantly displaced by other simpler and more convenient approaches, including the thermocouple and the resistance temperature detector (RTD). The electromotive force (EMF) principle is used in thermocouples, and electrical resistance changes are used in resistance temperature detectors (RTDs). Thermal radiation of hot bodies has served as the basis for radiation thermometers (once commonly referred to as radiation pyrometers and now called infrared thermometers) and has also been known and practiced for many decades.

Like all other areas of measurement, there is no single technology that is best for all applications. Each temperature-measurement technique has its own strengths and weaknesses. One responsibility of the instrument technician is to know these pros and cons so as to choose the best technology for the application.

【译文】

2.1 简介

温度是比热能更容易检测的量，当需要测量热能时，我们可以通过测量温度，然后根据热力学定律推断出所需测量的热能值。温度是化工、石油、石化、聚合物、塑料以及冶金和食品等流程工业中最为重要的过程变量。温度控制对于化学反应和物料分离（如蒸馏、干燥、蒸发、吸收、结晶、烘烤和挤压）等过程和操作至关重要。同时，温度控制在这些过程的安全运行中也起着至关重要的作用。

最常用的温度测量单位是华氏温标和摄氏温标。华氏温标是由丹尼尔·加布里埃尔·华伦海特发明的，并于1724年公布。尽管一些行业正在逐渐改用摄氏度，但它在美国仍被广泛使用。摄氏温标是瑞典科学家安德斯·摄尔修斯在1742年公布的，是使用最广泛的。

华氏度（°F）、摄氏度（℃）和开尔文（K，主要用于科学工作）是国际公认的测量温度的单位。实验确定的最低温度为－273.15℃。开氏温度标度的零值为－273.15℃，一个开氏温度单位的大小与摄氏温度相同。这些标尺之间的转换遵循以下方程：

$$℉=\left(℃\times\frac{9}{5}\right)+32 \tag{2-1}$$

$$℃=K-273.15 \tag{2-2}$$

温度是过程控制中最常见的过程变量。由于温度的测量范围很广（从绝对零度到数千度），有的温度的变化范围只有几度，灵敏度低到几分之一度，因此可以用于温度测量的设备范围也很广。

随温度变化的物理特性经常被用来测量温度。例如，加热时材料膨胀的特性可用于玻璃温度计、双金属温度计和温包系统。早期的温度计依赖于气体和液体的体积随温度的变化而变化，当然，这一原理在气液热系统以及液柱温度计中仍在应用。虽然这些温度仪表被广泛应用了许多年，但它们已被其他更简单方便的测量方法所取代，例如热电偶和热电阻温度计，热电势原理用于热电偶，电阻变化用于热电阻温度计。利用发热体的热辐射进行测量的是辐射温度计（以前通常称为辐射高温计，现在称为红外温度计），也已经被应用了几十年。

温度测量和其他测量领域一样，没有一种适合所有应用的技术，每种温度测量技术都有其优点和缺点。仪表技术人员应了解这些利弊，以便为相关应用选择最佳的测量方法。

2.2 Bimetallic Thermometers

Bonding two dissimilar metals with different coefficients of expansion produces a bimetallic strip when it is heated, it will bend and deform. The amount of deformation depends on the temperature, see Figure 2-1. These are used in bimetallic thermometers, temperature switches, and thermostats having a range of −50 to 500℃. When manufactured as a helix or coil, its movement with a change in temperature can move a pointer over a dial scale to indicate temperature. Industrial bimetallics use a helical coil to fit inside a stem. Most temperature switches operate on this principle, except that the pointer is replaced with a microswitch.

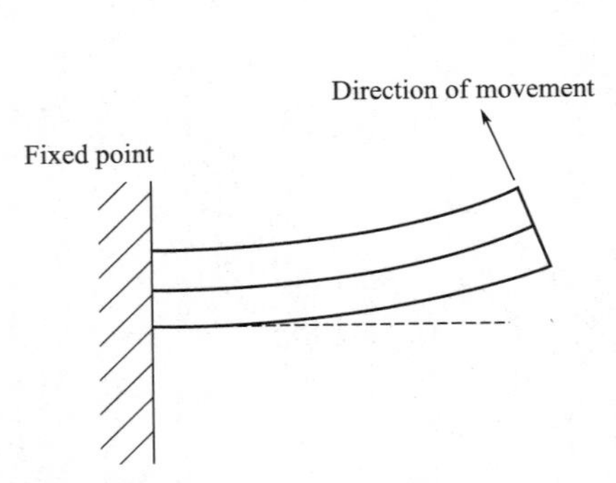

Figure 2-1 Bimetallic strip bent after heated

Rotating shaft
Free end attached to pointer shaft
Bulb
Fixed end

Figure 2-2 Bimetallic thermometer

Bimetallic thermometer (Figure 2-2) in general is very rugged and require little maintenance. It is a low-cost instrument, but has the disadvantages of relative inaccuracy and a relatively slow response time. It is normally used in temperature measurement applications that do not require high accuracy, such as measuring the temperature of pumps and bearings.

【译文】

2.2 双金属温度计

将具有不同膨胀系数的两种金属黏合在一起就可以制成双金属片，它受热时会产生弯曲形变，形变量的大小与温度有关，见图 2-1，它们用于制造温度范围为−50～500℃的双金属温度计、温度开关和恒温器。当双金属片被制造成螺旋或线圈形状时，它随温度变化的运动可以通过刻度盘上的指针移动来指示。工业用的双金属采用螺旋线圈并安装在阀杆内。大多数温度开关都是根据这一原理工作的，只是将指针换成了微开关。

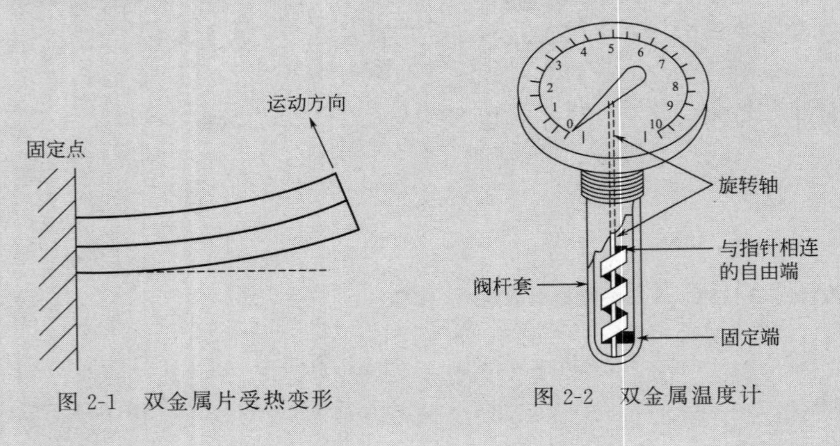

图 2-1 双金属片受热变形

图 2-2 双金属温度计

通常，双金属温度计（图 2-2）非常坚固，几乎不需要维护。它是一种低成本的仪表，但是存在精度不高、响应时间相对较慢的缺点。它通常用于不需要高精度的温度测量应用中，例如测量泵和轴承的温度。

2.3 Thermocouples

For many years the thermocouple (TC) was the first choice of instrumentation and control engineers in the process industries, but in recent years the position of the thermocouple has been increasingly challenged by the RTD. Nevertheless, the thermocouple still is used widely.

Thermocouples cover a range of temperatures, from −262 to+2760 ℃ and are manufactured in many materials, are relatively cheap, have many physical forms, all of which make them a highly versatile device.

A thermocouple can be considered as a heat-operated battery, which consists of two different types of homogeneous metal or alloy wires connected at one end. In order to use a thermocouple to measure process temperature, one end of the thermocouple has to be kept in contact with the process while the other end has to be kept at a constant temperature. The end that is in contact with the process is called the hot or measurement junction. The one that is kept at constant temperature is called cold or reference junction (see Figure 2-3).

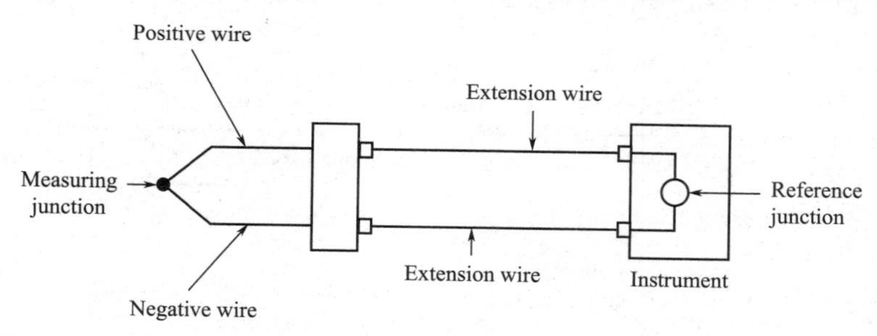

Figure 2-3 System wiring for thermocouple temperature measurement

A thermocouple current cannot be sustained in a circuit of a single homogeneous material, however varying in cross section, by the application of heat alone. The algebraic sum of the thermo electromotive forces (thermal EMFs) in a circuit composed of any number of dissimilar materials is zero if all of the circuit is at a uniform temperature. This means that a third homogeneous material always can be added to a circuit with no effect on the net EMF of the circuit so long as its extremities are at the same temperature. Therefore a device for measuring the thermal EMF may be introduced into a circuit at any point without affecting the resultant EMF, provided all the junctions added to the circuit are at the same temperature. It also follows that any junction whose temperature is uniform and that makes a good electrical contact does not affect the EMF of the thermocouple circuit regardless of the method used in forming the junction.

The structure of the thermocouple is shown in Figure 2-4. Terminal block is made of insulating material and used to support and join termination of wires. Connection head is a housing that encloses the terminal block and usually is provided with threaded openings for attachment to a protection tube and for attachment of a conduit. Connection head extension usually is a threaded fitting or an assembly of fittings extending between the thermowell or angle fitting and the connection head. Exact configuration depends on installation requirements. Protecting tube or well is used to protect thermocouple from damaging environmental effects.

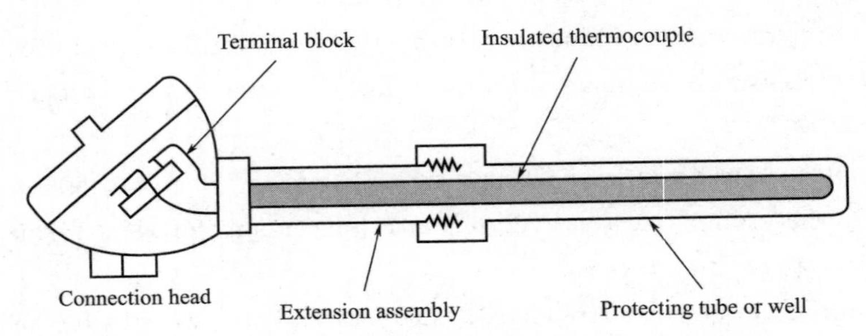

Figure 2-4 Structure of thermocouple

Theoretically, any two dissimilar metals will form a thermocouple. However, only a few are used because of their superior response to temperature changes (i. e. , sensitivity) and performance in general. There are many types of thermocouples, each with its advantages and disadvantages. Table 2-1 showing the more common thermocouple types and their standardized colors.

Table 2-1 Common thermocouple types

Type	Positive wire	Negative wire	Temperature range/℃
T	Copper(blue)	Constantan(red)	−200 to 350
J	Iron(white)	Constantan(red)	0 to 750
E	Chromel(violet)	Constantan(red)	0 to 900
K	Chromel(yellow)	Alumel(red)	0 to 1250
S	Pt90% -Rh10%(black)	Platinum(red)	0 to 1450

Copper-constantan (type T) can be used in either oxidizing or reducing atmospheres and its recommended temperature range is −200 to 350 ℃, this type exhibit a high resistance to corrosion from moisture, provide a relatively linear EMF output, and are good from the medium to the very low temperature range.

Iron-constantan (type J) can be used in reducing atmospheres. Type J thermocouple provide a very nearly linear EMF output, It is the least expensive commercially available type and its recommended temperature range is 0 to 750 ℃.

Chromel-constantan (type E) TCs provide the highest EMF per degree of temperature change, its recommended temperature range is 0 to 900 ℃. However, it also tends to drift more than the others. It can be used in oxidizing atmospheres.

Chromel-alumel (type K) can be used in oxidizing atmospheres and its recommended temperature range is 0 to 1250 ℃. It is the most linear TC in general use.

The platinum-platinum 90%/rhodium 10% (type S) TC is most important. It is used for high-accuracy, high-temperature applications. Industrial thermocouples of this material will match the standard calibration curve to better than ±0.25%. Its recommended temperature range is 0 to 1450 ℃.

If circuit EMF and reference EMF are known, measurement EMF can be calculated and the relative temperature determined. For each type of thermocouple there is a corresponding temperature-EMF reference table (i. e., Thermocouple tables) that converts mV reading into a temperature. Tables 2-2 to Tables 2-5 provide temperature vs. millivolts data for types S, K, E, and T thermocouples. All thermocouple tables are based upon a reference junction temperature of 0℃; therefore, direct conversion from the tables can be made only when an ice bath is used at the reference junction.

Table 2-2 Type S temperature-EMF table

℃	0	10	20	30	40	50	60	70	80	90
	Thermoelectric voltage in millivolts									
0	0.000	0.055	0.113	0.173	0.235	0.299	0.365	0.432	0.502	0.573
100	0.645	0.719	0.795	0.872	0.950	1.029	1.109	1.190	1.273	1.356
200	1.440	1.525	1.611	1.698	1.785	1.873	1.962	2.051	2.141	2.232
300	2.323	2.414	2.506	2.599	2.692	2.786	2.880	2.974	3.069	3.164
400	3.260	3.356	3.452	3.549	3.645	3.743	3.840	3.938	4.036	4.135
500	4.234	4.333	4.432	4.532	4.632	4.732	4.832	4.933	5.034	5.136
600	5.237	5.339	5.442	5.544	5.648	5.751	5.855	5.960	6.065	6.169
700	6.274	6.380	6.486	6.592	6.699	6.805	6.913	7.020	7.128	7.236
800	7.345	7.454	7.563	7.672	7.782	7.892	8.003	8.114	8.255	8.336
900	8.448	8.560	8.673	8.786	8.899	9.012	9.126	9.240	9.355	9.470
1000	9.585	9.700	9.816	9.932	10.048	10.165	10.282	10.400	10.517	10.635
1100	10.754	10.872	10.991	11.110	11.229	11.348	11.467	11.587	11.707	11.827
1200	11.947	12.067	12.188	12.308	12.429	12.550	12.671	12.792	12.912	13.034
1300	13.155	13.397	13.397	13.519	13.640	13.761	13.883	14.004	14.125	14.247
1400	14.368	14.610	14.610	14.731	14.852	14.973	15.094	15.215	15.336	15.456
1500	15.576	15.697	15.817	15.937	16.057	16.176	16.296	16.415	16.534	16.653
1600	16.771	16.890	17.008	17.125	17.243	17.360	17.477	17.594	17.711	17.826
1700	17.942	18.056	18.170	18.282	18.394	18.504	18.612	—	—	—

Table 2-3 Type K temperature-EMF table

℃	0	10	20	30	40	50	60	70	80	90
	Thermoelectric voltage in millivolts									
0	0. 000	0. 397	0. 798	1. 203	1. 611	2. 022	2. 436	2. 850	3. 266	3. 681
100	4. 095	4. 508	4. 919	5. 327	5. 733	6. 137	6. 539	6. 939	7. 338	7. 737
200	8. 137	8. 537	8. 938	9. 341	9. 745	10. 151	10. 560	10. 969	11. 381	11. 793
300	12. 207	12. 623	13. 039	13. 456	13. 874	14. 292	14. 712	15. 132	15. 552	15. 974
400	16. 395	16. 818	17. 241	17. 664	18. 088	18. 513	18. 938	19. 363	19. 788	20. 214
500	20. 640	21. 066	21. 493	21. 919	22. 346	22. 772	23. 198	23. 624	24. 050	24. 476
600	24. 902	25. 327	25. 751	26. 176	26. 599	27. 022	27. 445	27. 867	28. 288	28. 709
700	29. 128	29. 547	29. 965	30. 383	30. 799	31. 214	31. 214	32. 042	32. 455	32. 866
800	33. 277	33. 686	34. 095	34. 502	34. 909	35. 314	35. 718	36. 121	36. 524	36. 925
900	37. 325	37. 724	38. 122	38. 915	38. 915	39. 310	39. 703	40. 096	40. 488	40. 879
1000	41. 269	41. 657	42. 045	42. 432	42. 817	43. 202	43. 585	43. 968	44. 349	44. 729
1100	45. 108	45. 486	45. 863	46. 238	46. 612	46. 985	47. 356	47. 726	48. 095	48. 462
1200	48. 828	49. 192	49. 555	49. 916	50. 276	50. 633	50. 990	51. 344	51. 697	52. 049
1300	52. 398	52. 747	53. 093	53. 439	53. 782	54. 125	54. 466	54. 807	—	—

Table 2-4 Type E temperature-EMF table

℃	0	10	20	30	40	50	60	70	80	90
	Thermoelectric voltage in millivolts									
0	0. 000	0. 591	1. 192	1. 801	2. 419	3. 047	3. 683	4. 329	4. 983	5. 646
100	6. 317	6. 996	7. 683	8. 377	9. 078	9. 787	10. 501	11. 222	11. 949	12. 681
200	13. 419	14. 161	14. 909	15. 661	16. 417	17. 178	17. 942	18. 710	19. 481	20. 256
300	21. 033	21. 814	22. 597	23. 383	24. 171	24. 961	25. 754	26. 549	27. 345	28. 143
400	28. 943	29. 744	30. 546	31. 350	32. 155	32. 960	33. 767	34. 574	35. 382	36. 190
500	36. 999	37. 808	38. 617	39. 426	40. 236	41. 045	41. 853	42. 662	43. 470	44. 278
600	45. 085	45. 891	46. 697	47. 502	48. 306	49. 109	49. 911	50. 713	51. 513	52. 312
700	53. 110	53. 907	54. 703	55. 498	56. 291	57. 083	57. 873	58. 663	59. 451	60. 237
800	61. 022	61. 806	62. 588	63. 368	64. 147	64. 924	65. 700	66. 473	67. 245	68. 015
900	68. 783	69. 549	70. 313	71. 075	71. 835	72. 593	73. 350	74. 104	74. 857	75. 608
1000	76. 358	—	—	—	—	—	—	—	—	—

Table 2-5 Type T temperature-EMF table

℃	0	10	20	30	40	50	60	70	80	90
	Thermoelectric voltage in millivolts									
−200	−5.603	—	—	—	—	—	—	—	—	—
−100	−3.378	−3.378	−3.923	−4.177	−4.419	−4.648	−4.865	−5.069	−5.261	−5.439
0	0.000	0.383	−0.757	−1.121	−1.475	−1.819	−2.152	−2.475	−2.788	−3.089
0	0.000	0.391	0.789	1.196	1.611	2.035	2.467	2.980	3.357	3.813
100	4.277	4.749	5.227	5.712	6.204	6.702	7.207	7.718	8.235	8.757
200	9.268	9.820	10.360	10.905	11.456	12.011	12.572	13.137	13.707	14.281
300	14.860	15.443	16.030	16.621	17.217	17.816	18.420	19.027	19.638	20.252
400	20.869	—	—	—	—	—	—	—	—	—

If it is not possible to maintain the reference junction temperature at 0℃, a correction factor must be applied to the millivolt values shown in the temperature-EMF tables. Note that the millivoltage produced by a given thermocouple is decreased when the temperature difference between the measuring junction and the reference junction is decreased. Correcting for reference junction temperatures other than 0℃ is described below.

Step 1. From the appropriate TC table, obtain the millivoltage (based upon a 0℃) corresponding to the actual temperature of the thermocouple reference junction.

Step 2. Add the value obtained in step 1 to the millivoltage read on the potentiometer.

Step 3. The corrected millivoltage may then be converted into terms of temperature directly from the same table.

For example, a potentiometer indicates a 18.565 mV when connected to a type K thermocouple, and it is desired to convert this value to its equivalent temperature. The actual TC reference junction temperature, as determined by an accurate mercury-in-glass thermometer, is 20℃.

From the type K (Table 2-3), 20℃ corresponds to 0.798 mV, based upon a 0℃ reference junction, adding this value to the potentiometer reading, $18.565+0.798=19.363$ mV, which is the corrected millivoltage based upon a 0℃ reference junction. Interpolating from type K (Table 2-3), 19.363 mV corresponds to 470℃.

For most industrial applications the thermocouple has been popular, because it is relatively inexpensive, can be produced in a variety of sizes, can be of ruggedized construction and covers a wide temperature range. Thermocouples are also small, convenient, and versatile (can be welded to a pipe), cover wide ranges, are reasonably stable, reproducible, accurate, and fast. The EMF they generate is independent of wire length and diameter. While RTDs are more accurate and more stable and while thermistors are more sensitive, thermocouples are the most economical and the best to detect the highest temperatures.

The main disadvantage of the thermocouple is its weak output signal. This makes it sensitive to electrical noise and limits its use to relatively wide spans (usually the minimum transmitter span is 1.0 mV). It is nonlinear, and the conversion of the EMF generated into temperature is not as simple as in direct reading devices. TCs always require amplifiers, and the calibration of the TC can change due to contamination or composition changes due to internal oxidation, cold-working or temperature gradients. Another limitation is that bare TCs cannot be used in conductive fluids, and if their wires are not homogeneous, this can cause errors.

The weakest link in virtually all measurements is the temperature sensor. In general, one should use the largest size TC wire possible, and avoid stress and vibration. Use of integral transmitters is also recommended whenever possible. In addition, one should avoid steep temperature gradients, and be careful in selecting the sheath and thermowell materials.

【译文】

2.3 热电偶

多年来，热电偶是仪表和控制工程师在流程工业中的首选，但近年来，热电偶的地位越来越受到热电阻的挑战。尽管如此，热电偶仍被广泛使用。

热电偶的测温范围从－262℃到＋2760℃。它可以由多种材料制造，相对便宜，有多种形式，所以成为一个通用测温装置。

热电偶可以被认为是一种热电池，它由两种不同类型的均质金属或合金导线连接在一个端点。为了使用热电偶测量工艺温度，热电偶的一端必须与工艺介质保持接触，另一端必须保持恒温。与工艺接触的一端称为热端或测量端，保持恒温的另一端称为冷端或参考端（图 2-3）。

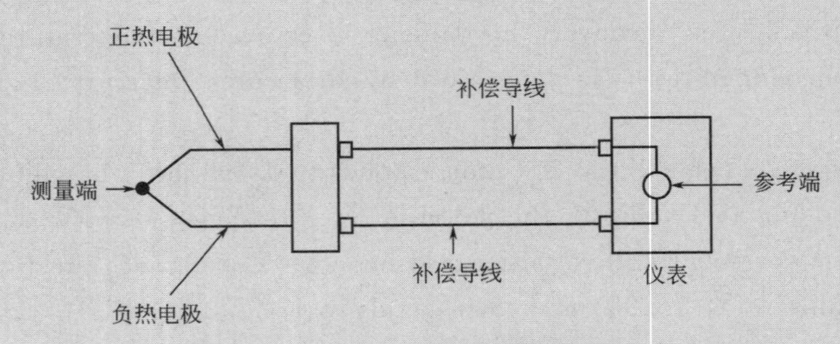

图 2-3　热电偶温度测量系统的连接

热电偶电流在单一均匀材料的电路中，不管其横截面如何变化，仅靠温差是无法维持的。由不同材料组成的电路，如果处于同一温度，则其热电势为零。这意味着，只要电路的两端温度相同，第三种均匀材料可以加入到电路中，而不会对电路的电动势产生影响。因此，可以在任何点将测量热电势的装置引入电路，而不影响热电偶产生的电势，前提是加入到电路中的所有接点都处于相同的温度。我们还可以得出结论，任何温度均匀且具有良好电接触的接点，都不会影响热电偶电路的电动势，无论形成该接点的方法如何。

热电偶的结构如图 2-4 所示。接线盒由绝缘材料制成，用于支撑和连接导线的端部。连接头是一个外壳，内部有接线盒，通常具有螺纹开口，用于连接保护套管和管线。连接头加长件通常是一个螺纹连接件，它们延伸在套管或角形连接件和连接头之间，具体的配置取决于安装要求。保护套管用于保护热电偶免受外部环境的影响。

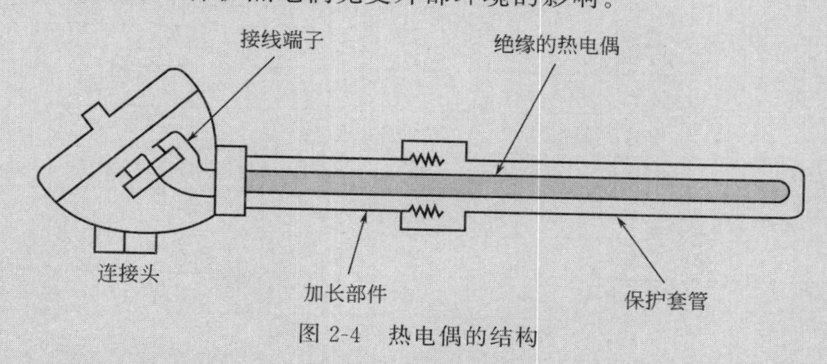

图 2-4　热电偶的结构

理论上，任何两种不同的金属都会形成热电偶。然而，由于这些金属对温度变化的反应（灵敏度）和总体性能的差异，只有少数一些材料能用以制造热电偶。热电偶有很多种，优缺点各异。表 2-1 所示为常见的热电偶类型及其标准颜色。

表 2-1　常见的热电偶类型

类型	正电极丝	负电极丝	温度范围/℃
T	铜(蓝色)	康铜(红色)	－200～350
J	铁(白色)	康铜(红色)	0～750
E	镍铬合金(紫色)	康铜(红色)	0～900
K	镍铬合金(黄色)	镍铝(红色)	0～1250
S	铂 90%-铑 10%(黑色)	铂(红色)	0～1450

铜-康铜（T 型）可用于氧化或还原气氛，其推荐的温度范围为－200～350℃。这种类型的热电偶具有较好的耐潮湿和耐腐蚀能力，能提供相对线性的电动势输出，并且低温范围内表现良好。

铁-康铜（J 型）可用于还原性气氛。J 型热电偶可提供一个非常接近线性的电动势输出，它是最便宜的商业可用的类型，其推荐的温度范围为 0～750℃。

铬镍-康铜（E 型）热电偶提供最高的温度与电动势变化率，其推荐的温度范围是 0～900℃，可用于氧化气氛。

镍铬（K 型）可用于氧化气氛，其推荐温度范围为 0～1250℃。它是常用的线性最好的热电偶。

铂-铂 90%/铑 10%（S 型）热电偶最为重要，它用于高精度和高温应用。该材料的工业热电偶与标准校准曲线的匹配度优于±0.25%。它的推荐温度范围是 0～1450℃。

如果已知电路热电势和参考热电势，则可以计算测量总热电势和确定被测温度。每一种热电偶都有相应的温度-电动势参考表（即热电偶分度表），它能把热电偶的毫伏读数转换成温度。表 2-2 到表 2-5 提供了 S 型、K 型、E 型和 T 型热电偶的分度表数据。所有的热电偶分度表数据均基于参考端（冷端）温度为 0℃而获得。因此，只有在参考端使用冰浴（0℃）时，才能根据表进行直接转换。

表 2-2 S 型热电偶分度表

℃	0	10	20	30	40	50	60	70	80	90
	热电动势/mV									
0	0.000	0.055	0.113	0.173	0.235	0.299	0.365	0.432	0.502	0.573
100	0.645	0.719	0.795	0.872	0.950	1.029	1.109	1.190	1.273	1.356
200	1.440	1.525	1.611	1.698	1.785	1.873	1.962	2.051	2.141	2.232
300	2.323	2.414	2.506	2.599	2.692	2.786	2.880	2.974	3.069	3.164
400	3.260	3.356	3.452	3.549	3.645	3.743	3.840	3.938	4.036	4.135
500	4.234	4.333	4.432	4.532	4.632	4.732	4.832	4.933	5.034	5.136
600	5.237	5.339	5.442	5.544	5.648	5.751	5.855	5.960	6.065	6.169
700	6.274	6.380	6.486	6.592	6.699	6.805	6.913	7.020	7.128	7.236
800	7.345	7.454	7.563	7.672	7.782	7.892	8.003	8.114	8.255	8.336
900	8.448	8.560	8.673	8.786	8.899	9.012	9.126	9.240	9.355	9.470
1000	9.585	9.700	9.816	9.932	10.048	10.165	10.282	10.400	10.517	10.635
1100	10.754	10.872	10.991	11.110	11.229	11.348	11.467	11.587	11.707	11.827
1200	11.947	12.067	12.188	12.308	12.429	12.550	12.671	12.792	12.912	13.034
1300	13.155	13.397	13.397	13.519	13.640	13.761	13.883	14.004	14.125	14.247
1400	14.368	14.610	14.610	14.731	14.852	14.973	15.094	15.215	15.336	15.456
1500	15.576	15.697	15.817	15.937	16.057	16.176	16.296	16.415	16.534	16.653
1600	16.771	16.890	17.008	17.125	17.243	17.360	17.477	17.594	17.711	17.826
1700	17.942	18.056	18.170	18.282	18.394	18.504	18.612	—	—	—

表 2-3 K 型热电偶分度表

℃	0	10	20	30	40	50	60	70	80	90
	热电动势/mV									
0	0.000	0.397	0.798	1.203	1.611	2.022	2.436	2.850	3.266	3.681
100	4.095	4.508	4.919	5.327	5.733	6.137	6.539	6.939	7.338	7.737
200	8.137	8.537	8.938	9.341	9.745	10.151	10.560	10.969	11.381	11.793
300	12.207	12.623	13.039	13.456	13.874	14.292	14.712	15.132	15.552	15.974
400	16.395	16.818	17.241	17.664	18.088	18.513	18.938	19.363	19.788	20.214
500	20.640	21.066	21.493	21.919	22.346	22.772	23.198	23.624	24.050	24.476
600	24.902	25.327	25.751	26.176	26.599	27.022	27.445	27.867	28.288	28.709
700	29.128	29.547	29.965	30.383	30.799	31.214	31.214	32.042	32.455	32.866
800	33.277	33.686	34.095	34.502	34.909	35.314	35.718	36.121	36.524	36.925
900	37.325	37.724	38.122	38.915	38.915	39.310	39.703	40.096	40.488	40.879
1000	41.269	41.657	42.045	42.432	42.817	43.202	43.585	43.968	44.349	44.729
1100	45.108	45.486	45.863	46.238	46.612	46.985	47.356	47.726	48.095	48.462
1200	48.828	49.192	49.555	49.916	50.276	50.633	50.990	51.344	51.697	52.049
1300	52.398	52.747	53.093	53.439	53.782	54.125	54.466	54.807	—	—

表 2-4　E 型热电偶分度表

℃	0	10	20	30	40	50	60	70	80	90
	热电动势/mV									
0	0.000	0.591	1.192	1.801	2.419	3.047	3.683	4.329	4.983	5.646
100	6.317	6.996	7.683	8.377	9.078	9.787	10.501	11.222	11.949	12.681
200	13.419	14.161	14.909	15.661	16.417	17.178	17.942	18.710	19.481	20.256
300	21.033	21.814	22.597	23.383	24.171	24.961	25.754	26.549	27.345	28.143
400	28.943	29.744	30.546	31.350	32.155	32.960	33.767	34.574	35.382	36.190
500	36.999	37.808	38.617	39.426	40.236	41.045	41.853	42.662	43.470	44.278
600	45.085	45.891	46.697	47.502	48.306	49.109	49.911	50.713	51.513	52.312
700	53.110	53.907	54.703	55.498	56.291	57.083	57.873	58.663	59.451	60.237
800	61.022	61.806	62.588	63.368	64.147	64.924	65.700	66.473	67.245	68.015
900	68.783	69.549	70.313	71.075	71.835	72.593	73.350	74.104	74.857	75.608
1000	76.358	—	—	—	—	—	—	—	—	—

表 2-5　T 型热电偶分度表

℃	0	10	20	30	40	50	60	70	80	90
	热电动势/mV									
−200	−5.603	—	—	—	—	—	—	—	—	—
−100	−3.378	−3.378	−3.923	−4.177	−4.419	−4.648	−4.865	−5.069	−5.261	−5.439
0	0.000	0.383	−0.757	−1.121	−1.475	−1.819	−2.152	−2.475	−2.788	−3.089
0	0.000	0.391	0.789	1.196	1.611	2.035	2.467	2.980	3.357	3.813
100	4.277	4.749	5.227	5.712	6.204	6.702	7.207	7.718	8.235	8.757
200	9.268	9.820	10.360	10.905	11.456	12.011	12.572	13.137	13.707	14.281
300	14.860	15.443	16.030	16.621	17.217	17.816	18.420	19.027	19.638	20.252
400	20.869	—	—	—	—	—	—	—	—	—

如果不能将参考端温度保持在 0℃，则必须对分度表中所示的毫伏值进行补偿校正。注意，当测量端和参考端之间的温差减小时，热电偶产生的毫伏电压就会减小。以下是对参考端温度非 0℃时的具体修正步骤。

第一步，从相应的分度表中，得到与热电偶参考端的实际温度相对应的毫伏电压（基于 0℃）。

第二步，将第一步中得到的值加到电位计的毫伏电压读数上。

第三步，将补偿修正后的毫伏电压，再次直接从同一分度表转换成正确的温度值。

例如，与 K 型热电偶配套连接的电位计，测得回路热电势为 18.565 mV，希望将该热电势值转换成测量端实际的对应温度。此时热电偶的参考端温度由一个精确的玻璃水银温度计测量是 20℃。

根据 K 型分度表，基于 0℃参考端，20℃对应 0.798 mV，把这个值加到电位计的读数上：18.565＋0.798 ＝ 19.363 mV，这是基于 0℃参考端的补偿修正毫伏电压。通过查询 K 型分度表，确定测量端实际温度为 470℃。

对于大多数工业应用来说，热电偶是很受欢迎的，因为它相对便宜，规格尺寸多，结构坚固，可以覆盖很宽的温度范围。热电偶体积小、使用方便、用途广泛（可以焊接到管道上）、范围广、稳定性好、重复性好、精度高、速度快。它们产生的电动势与导线的长度和直径无关。相比而言，虽然热电阻更准确和稳定，更灵敏，但热电偶仍然是最经济和最适合高温检测的。

热电偶的主要缺点是输出信号弱，这使得它对电噪声很敏感，并限制了它的测温范围（通常变送器最小量程为 1.0mV），而且热电势是非线性的，导致了热电偶产生的电动势转换成温度不像直接读数那样简单。热电偶变送器需要放大器，而且由于内部氧化和温度梯度等因素影响，它的标定可能会改变。另一个限制是裸露的热电偶不能用于导电流体测温，如果它们的导线不均匀，就会造成误差。

在所有实际测量中，最薄弱的环节是温度传感器。一般情况下，应尽量使用最大尺寸的热电极线，并应避免应力和振动。如果可能的话，建议使用整体型变送器。此外，应避免大的温度梯度，并恰当选择保护套管和热电偶材料。

2.4 Resistance Temperature Detectors

For most metals the change in electrical resistance is directly proportional to its change in temperature and is linear over a range of temperatures. This constant factor called the temperature coefficient of electrical resistance is the basis of resistance temperature detectors (RTDs). The RTD can actually be regarded as a high precision wire wound resistor whose resistance varies with temperature. By measuring the resistance of the metal, its temperature can be determined. The electrical resistance of RTD changes with temperature as approximated by the following formula:

$$R_T = R_{ref}[1+\alpha(T-T_{ref})] \tag{2-3}$$

Where

R_T = Resistance of RTD at given temperature T (Ω)

R_{ref} = Resistance of RTD at the reference temperature T_{ref} (Ω)

α = Temperature coeffcient of resistance (℃$^{-1}$)

Only a few pure metals have a characteristic relationship suitable for the fabrication of sensing elements used in RTDs. The metal must have an extremely stable resistance-temperature relationship so that neither the absolute value of the resistance R_{ref} nor the coefficient α drift with repeated heating and cooling within the thermometer's specified temperature range of operation. The material must exhibit relatively small resistance changes for nontemperature effects, such as strain and possible contamination which may not be totally eliminated from a controlled manufacturing environment. The material's change in resistance with temperature must be relatively large in order to produce a resultant thermometer with inherent sensitivity. The metal must not undergo any change of phase or state within a reasonable temperature range. Finally, the metal must be commercially available with essentially a consistent resistance-temperature relationship to provide reliable uniformity.

Several different pure metals (such as platinum, nickel and copper) can be used in the manufacture of an RTD. The entire resistance thermometer is an assembly of parts, which include the sensing element, internal leadwires, internal supporting and insulating materials, and protection tube or case.

Platinum RTDs Of all materials currently utilized in the fabrication of thermoresistive elements, platinum has the optimum characteristics for service over a wide temperature range. Although platinum is a noble metal and does not oxidize, it is subject to contamination at elevated temperatures by some gases, such as carbon monoxide and other reducing atmospheres, and by metallic oxides.

Nickel RTDs Nickel RTDs are often used in situations where the temperature range is small and the sensitivity is high. Current availability of nickel sensors has continued primarily as a component replacement for already existing industrial systems.

Copper RTDs Copper RTDs are inexpensive, and are generally used in places where the accuracy of the measurement is not high and the temperature is low. The straight-line characteristics of copper have in the past been useful in allowing two sensors to be applied directly for temperature-difference measurements.

RTDs are usually protected from the environment by a sheath made of stainless steel or any other temperature-and corrosion-resistant material (see Figure 2-5). The element fits snugly inside the sheath to produce a high rate of heat transfer. Ceramic insulators are typically used to isolate the internal lead wires. At the end of the tube a hermetic seal protects the element. The assembly may be terminated with the lead wires or may be supplied with an appropriate terminal block similar to a T/C assembly.

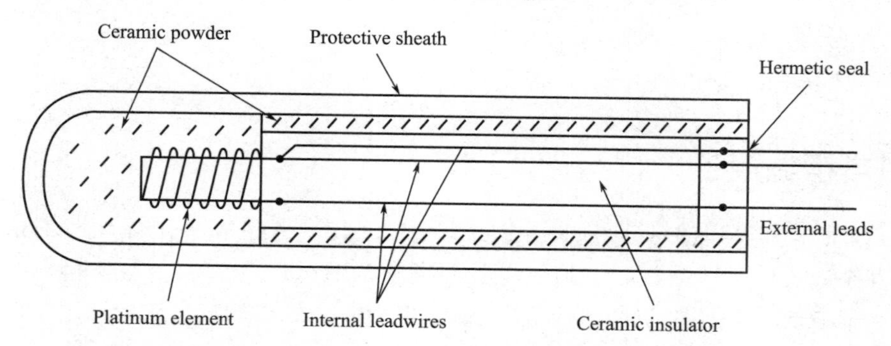

Figure 2-5 Typical RTD and thermowell construction

By measuring the resistance of the RTD element one can determine the process temperature if the change in total resistance measured is affected by nothing but the process temperature. The RTD element is connected by two or three lead wires to the readout or transmitting instrument.

With a two-wired method (see Figure 2-6), the effect of the lead wire resistance (and the effects of the change in resistance that occurs as the ambient temperature changes) introduces significant errors.

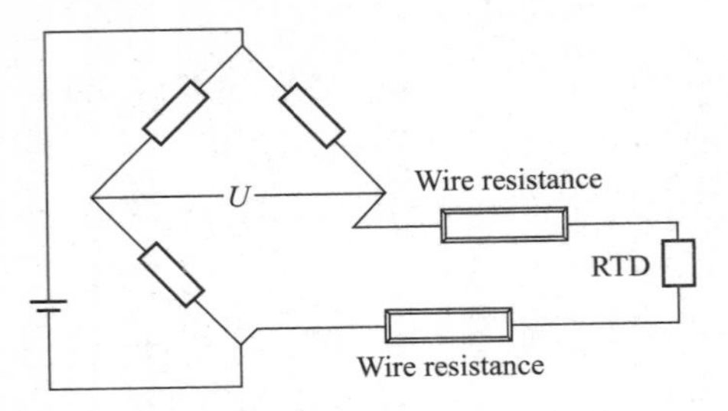

Figure 2-6　Two-wired method

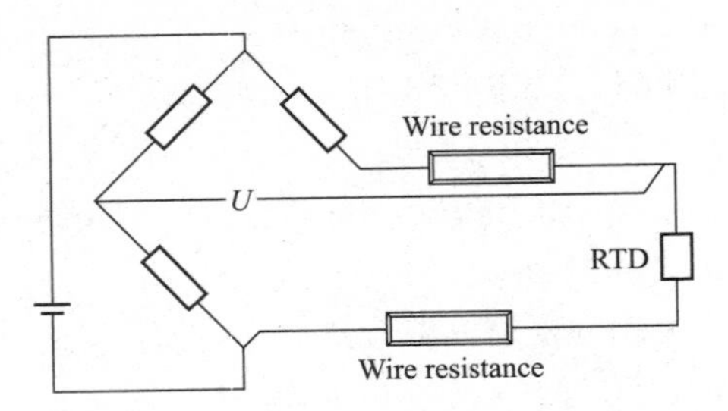

Figure 2-7　Three-wired method

With a three-wired method (see Figure 2-7), the impedance in the wires will cancel because the wires are in opposite legs of the bridge. In other words, the three-wire method compensates for the effect of lead resistance. This is the most practical and commonly used RTD method.

Of all temperature-measuring devices, RTDs are, at moderate temperatures, the most stable and the most accurate. Their output is stronger than that of a T/C, they are less susceptible to electrical noise, and they operate on a higher level of electrical signals. Moreover, they are more sensitive and more linear than a T/C (output versus temperature), use copper extension wire (not special extension wire), require no reference junction, and are easy to interchange.

However, RTDs are relatively expensive compared to thermocouples, have a slow response, and require a current source. They are susceptible to small resistance changes, and self-heating appears as a measurement error (the main source of RTD error). In addition, RTDs have a limited temperature range, are susceptible to strain and vibration, generate some nonlinearity, and require three extension wires. Their resistance curves vary from manufacturer to manufacturer, and their accuracy and service life are limited at high temperatures.

【译文】

2.4　热电阻温度计

对于大多数金属，电阻的变化与温度的变化成正比，并且在一定温度范围内呈线性关系，该常数称为电阻温度系数，这是热电阻温度检测器（RTD）的检测基础。实际上，热电阻可以看作是电阻值随温度变化的高精度绕线电阻器。通过测量金属的电阻，就可以确定其温度。热电阻的电阻随温度变化，如下式所示。

$$R_T = R_{ref}[1+\alpha(T-T_{ref})] \tag{2-3}$$

式中，R_T 是实际温度 T 时的电阻值，Ω；R_{ref} 是参考温度 T_{ref} 时的阻值，Ω；α 是电阻温度系数，℃$^{-1}$。

只有少数几种纯金属具有适于制作热电阻温度计传感元件的特性关系。这种金属必须具有一种非常稳定的电阻与温度的对应关系，使电阻 R_{ref} 的绝对值和系数 α 在反复加热和冷却的情况下，都不会在温度计规定的工作温度范围内漂移。对于无法从可控的制造环境中完全消除的一些非温度效应，如应变和可能的污染，材料必须表现出相对较小的电阻变化。为了得到具有固有灵敏度的温度计，材料的电阻值随温度的变化必须相当大。在合理的温度范围内，金属不得发生任何相或态的变化。最后，这种电阻与温度关系具有一致性的金属必须很容易获得。

几种纯金属（如铂、镍和铜）可用于制造热电阻温度计。整个电阻温度计是一个部件的组装，包括传感元件、内部引线、内部支撑和绝缘材料以及保护管或外壳。

铂电阻温度计 在目前用于制造热电阻元件的所有材料中，金属铂具有在较大的温度范围内的最佳特性。虽然铂是一种贵金属，不易氧化，但在高温下，它会受到一些气体的污染，如一氧化碳和其他还原性气体，还会受金属氧化物的污染。

镍电阻温度计 镍电阻温度计常用于温度变化范围小、灵敏度要求较高的场合。目前镍传感器主要是作为现有工业系统的组件替换。

铜电阻温度计 铜电阻价格低廉，一般用于测量准确度要求不高且温度较低的场合。铜电阻的直线特性在允许两个传感器直接应用于温差测量时是有用的。

热电阻温度计通常由不锈钢或其他耐高温耐腐蚀材料制成的套管加以保护，不受环境影响（图 2-5）。热电阻元件紧贴套管以获得较高的传热率。陶瓷绝缘体通常用于隔离内部引线。在套管的末端有一个密封装置保护电阻元件。该总成可以用引线或用类似于热电偶里的端子块引出信号。

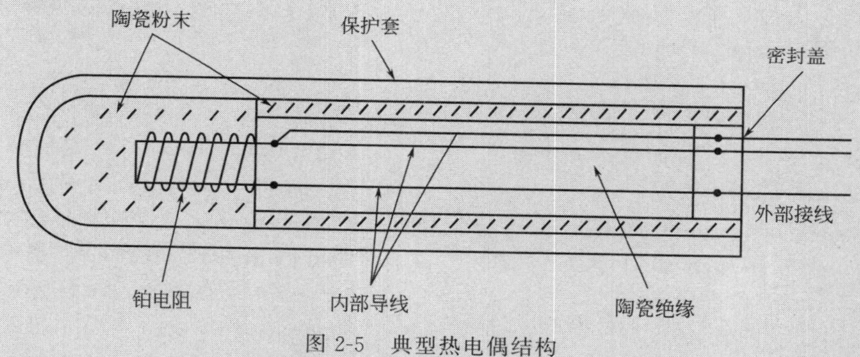

图 2-5 典型热电偶结构

如果热电阻的变化只受温度的影响，那么我们可以通过测量电阻值的变化来确定温度值。在实际热电阻温度计安装中，电阻元件由两线制或三线制导线连接到指示仪表或变送器上。

对于两线制接线（图 2-6），引线电阻的影响（以及环境温度变化引起的电阻变化的影响）引入了显著的误差。

对于三线制接线（图 2-7），由于导线在电桥的相对支路中，导线中的阻抗将会抵消。换句话说，三线法补偿了引线电阻的影响。这是最常用的安装方法。

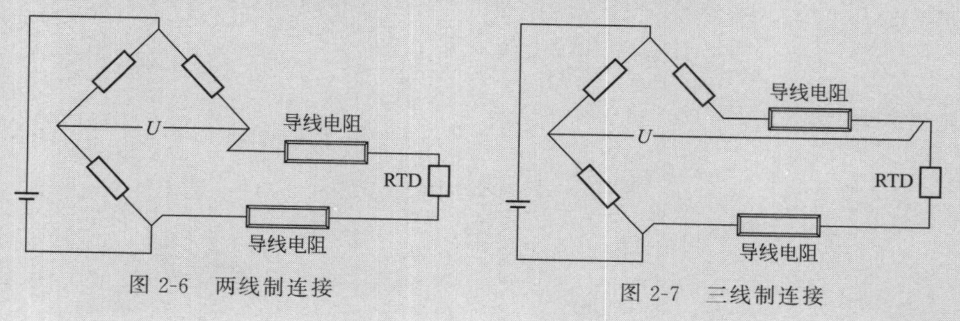

图 2-6 两线制连接

图 2-7 三线制连接

在所有温度测量装置中，热电阻温度计在中等温度下是最稳定和最准确的。它们的输出比热电偶强，对电噪声的敏感性较低，并且电信号较高。此外，它们比热电偶（输出随温度的变化）更灵敏、更线性，使用一般的铜线作延伸线而不需要特殊的延伸线，也不需要参考端，且易于更换。

然而，与热电偶相比，热电阻温度计相对昂贵，响应速度慢，需要电流源。它们容易受到小的电阻变化的影响，并且电阻元件的自热会带来测量误差（热电阻温度计的主要误差来源）。此外，热电阻温度计的温度范围有限，易受应变和振动的影响，也会产生一些非线性，需要三根延长线。它们的电阻曲线因制造商而异，在高温下其精度和使用寿命有限。

Words and Expressions 词汇和短语

1. thermodynamic /ˌθɜː rmoʊdaɪˈnæmɪk/ adj. 热力学的
2. chemical /ˈkemɪkl/ adj. 化工的
3. petroleum /pəˈtroʊliəm/ n. 石油
4. petrochemical /ˌpetroʊˈkemɪkl/ adj. 石化的
5. polymer /ˈpɑːlɪmər/ n. 聚合物
6. plastic /ˈplæstɪk/ n. 塑料
7. metallurgical /ˌmetlˈɜːrdʒɪkl/ adj. 冶金的
8. reaction /riˈækʃn/ n. 反应
9. distillation /ˌdɪstɪˈleɪʃn/ n. 精馏
10. evaporation /ɪˌvæpəˈreɪʃn/ n. 蒸发
11. Fahrenheit scale 华氏温标
12. Celsius scale 摄氏温标
13. liquid-in-glass 玻璃温度计
14. bimetallic /ˌbaɪməˈtælɪk/ adj. 双金属的
15. filled-system 温包系统
16. resistance temperature detector 热电阻温度计
17. electromotive force 电动势
18. thermal radiation 热辐射
19. infrared thermometer 红外测温仪
20. thermostat /ˈθɜːrməstæt/ n. 恒温器
21. pointer /ˈpɔɪntər/ n. 指针
22. response time 响应时间
23. homogeneous /ˌhoʊməˈdʒiːniəs/ adj. 均匀的
24. measurement junction 测量端
25. reference junction 参考端
26. terminal block 接线盒
27. threaded /ˈθredɪd/ adj. 螺纹的
28. assembly /əˈsembli/ n. 装配，组合
29. positive wire （热电偶）正电极丝
30. negative wire （热电偶）负电极丝
31. oxidize /ˈɑːksɪdaɪz/ vt. 使氧化
32. reducing /rɪˈdusɪŋ/ adj. 还原性的
33. correction factor 校正因子
34. potentiometer /pəˌtenʃiˈɑːmɪtər/ n. 电位计
35. stress /stres/ n. 应力
36. vibration /vaɪˈbreɪʃn/ n. 振动
37. platinum /ˈplætɪnəm/ n. 铂
38. nickel /ˈnɪkl/ n. 镍
39. copper /ˈkɔpə/ n. 铜
40. protection tube 防护套管

Chapter 3

Pressure Measurement

第3章

压力测量

3.1 Introduction

Pressure is defined as force per unit area and may be expressed in units of newtons per square meter, millimeters of mercury, atmospheres. Pressure is one of the most commonly used and important measurement in process control. Many types of industrial measurements are actually inferred from pressure, such as flow, level, weight and liquid density.

The most familiar pressure measuring devices are manometers and dialgauges, but these require a manual operator.

For use in process control, a pressure measuring device needs a pressure transmitter that will produce an output signal for transmission, e. g. an electric current proportional to the pressure being measured. A transmitter typically that produces an output of a 4 to 20 mA signal is rugged and can be used in flammable or hazardous service.

Pressure can be expressed in three forms (see Figure 3-1):

- Absolute pressure, where the reference is complete vacuum
- Gage pressure, where the reference is atmospheric pressure
- Differential pressure, which represents the difference between two pressure levels (note that gage pressure is a differential pressure between a value and atmospheric pressure).

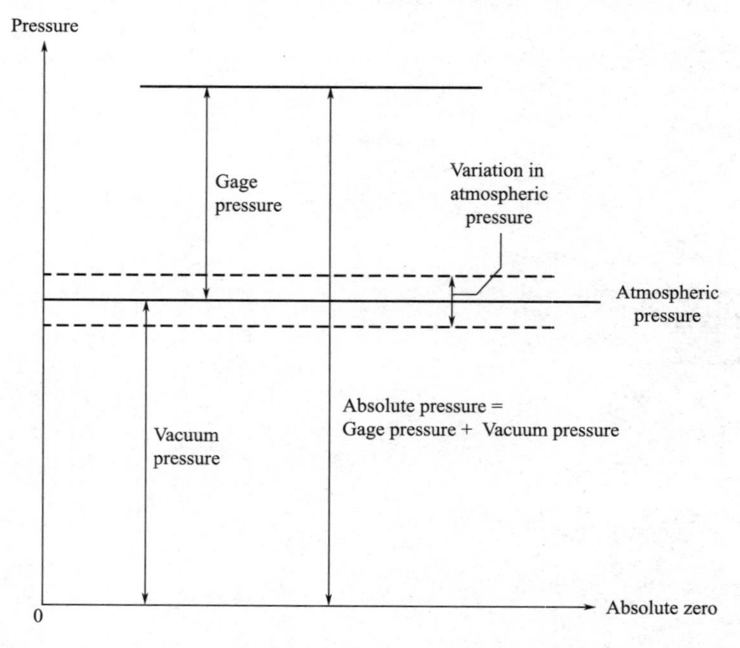

Figure 3-1 Absolute, gage and differential pressure

【译文】

3.1 简介

压强定义为单位面积上的力，可以用牛顿每平方米、毫米汞柱、大气压等单位来表示。

压力是过程控制中最常用和最重要的测量变量之一，可以从压力测量中推断出许多类型的工业变量值，例如流量、液位、重量和液体密度。

最常见的压力测量装置是液压计和千分表，但这些装置需要人工操作。

在过程控制中，压力测量装置需要变送器来产生一个输出信号进行传输，例如与被测压力成比例的电流信号。典型的变送器输出可靠的4～20mA信号，可用于易燃易爆或危险的场合。

压力有三种表示形式（图3-1）：

• 绝对压力，以真空为参考压力；

• 表压，以大气压为参考压力；

• 差压，表示两个压力之间的差异（注意，表压也是一种差压，它是与大气压之间的差压）。

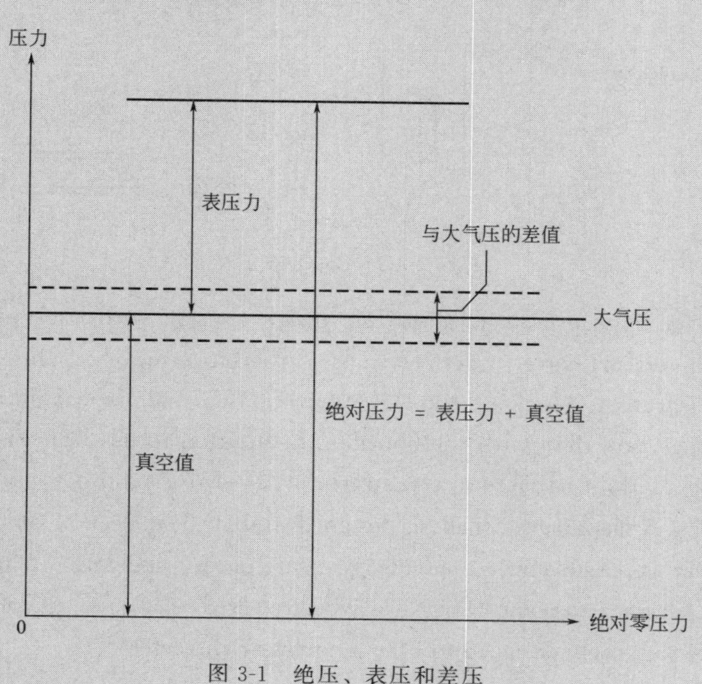

图3-1　绝压、表压和差压

3.2 Mechanical Pressure Elements

One class of pressure sensors uses some form of elastic element whose geometry is altered by changes in pressure. This mechanical movement is converted, through gears and pivots, into a pointer on a graduated dial. These elements are of three principal types: bourdon tube, diaphragm, and bellows.

Bourdon tube (Figure 3-2) is a curved or twisted tube whose transfer section is generally oval. In principle, it is a tube closed at one end, with an internal cross section that is not a perfect circle and, if bent or distorted, has the property of changing its shape with internal pressure variations. One end of the tube is linked to the process pressure, and the other end is sealed and linked to the mechanism operating the pointer. As the pressure increases, the tube tends to straighten itself out. This movement is indicated by the pointer.

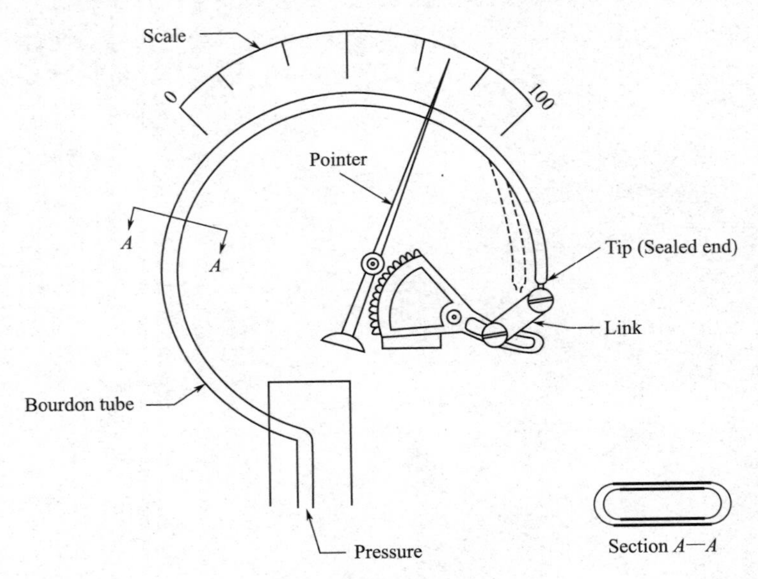

Figure 3-2 Bourdon tube

Diaphragm (see Figure 3-3) is a flexible disk, usually with concentric corrugations, that is used to convert pressure to deflection. It is therefore more capable of measuring low pressures and less prone to blockage. On the other hand, if the diaphragm ruptures there is a much higher leakage rate than with the bourdon. In addition to use in pressure sensors, diaphragms can serve as fluid barriers in transmitters, as seal assemblies, and also as building blocks for capsules. A diaphragm usually is designed so that the deflection-versus-pressure characteristics are as linear as possible over a specified pressure range, and with a minimum of hysteresis and minimum shift in the zero point. However, when required, as in the case of an altitude sensor, a diaphragm can be purposely designed to have a nonlinear characteristic.

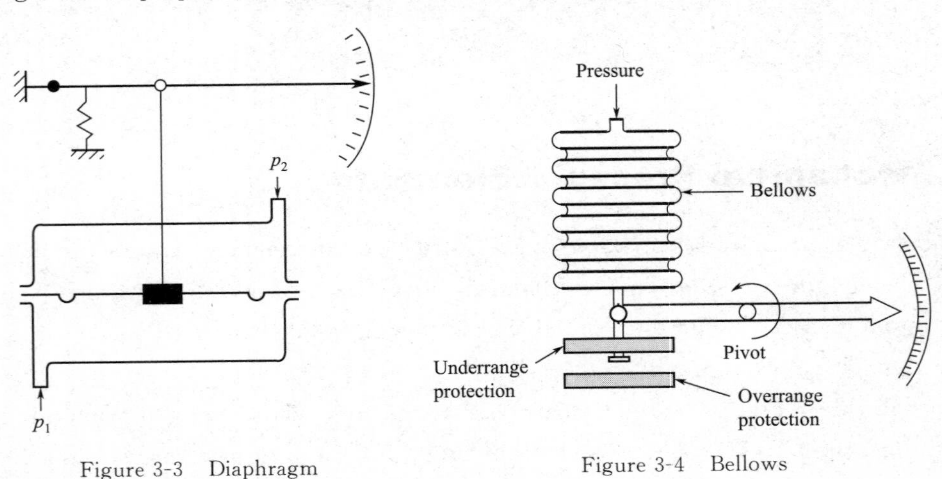

Figure 3-3 Diaphragm

Figure 3-4 Bellows

Bellows (Figure 3-4) is an expandable element and is made up of a series of folds, which allow expansion. One end of the bellows is fixed and the other moves in response to the applied pressure. Most bellows are made from seamless tubes, and the convolutions either are hydraulically formed or mechanically rolled. Bellows elements are well adapted to use in applications that require long strokes. They are well suited for input elements for large-case recorders and indicators and for feedback elements in pneumatic controllers.

【译文】

3.2 机械压力元件

一类压力传感器使用某种形式的弹性元件，它的几何形状会随着压力的变化而改变。这种机械运动通过齿轮和枢轴转换成刻度盘上的指针。这些弹性元件主要有三种类型：弹簧管、膜片和波纹管。

弹簧管（图 3-2）是一种弯曲或扭曲的管，其传输截面一般是椭圆形。从原理上讲，它是一个一端封闭的管子，由于内部截面不是一个完美的圆，当受压弯曲或扭曲时，其形状会随着内部压力的变化而改变。弹簧管的一端与过程压力相连，另一端密封并与指针机构相连。随着压力的增加，管子会逐渐变直，并由指针指示出对应的过程压力。

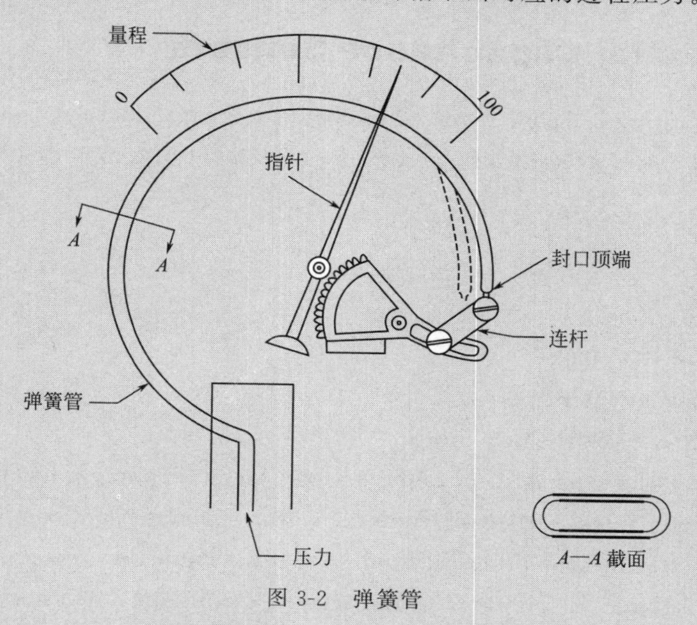

图 3-2 弹簧管

膜片（图 3-3）是一种柔性磁盘，通常具有同心波纹，用于将压力转换为挠度。因此，它更能测量低压，且不容易堵塞。但另一方面，如果膜片破裂，它的泄漏率比弹簧管要高得多。除了用于压力传感器外，膜片还可以用作变送器中的流体隔离和密封组件，也可以用作膜盒的构件。膜片的设计通常使挠度和压力之间的特性尽可能线性，并且在零点处具有最小的滞后和最小的偏移。此外，某些情况下，如在高度传感器中，膜片可以设计成非线性的特性。

波纹管（图 3-4）是一种可膨胀的元件，由一系列可膨胀的褶皱组成。波纹管的一端固定，另一端根据施加的压力移动。大多数波纹管是由无缝管制成的，通过液压成形或机械轧制。波纹管元件适用于长冲程的应用，常用于大容量记录仪和指示器的输入元件和气动控制器的反馈元件。

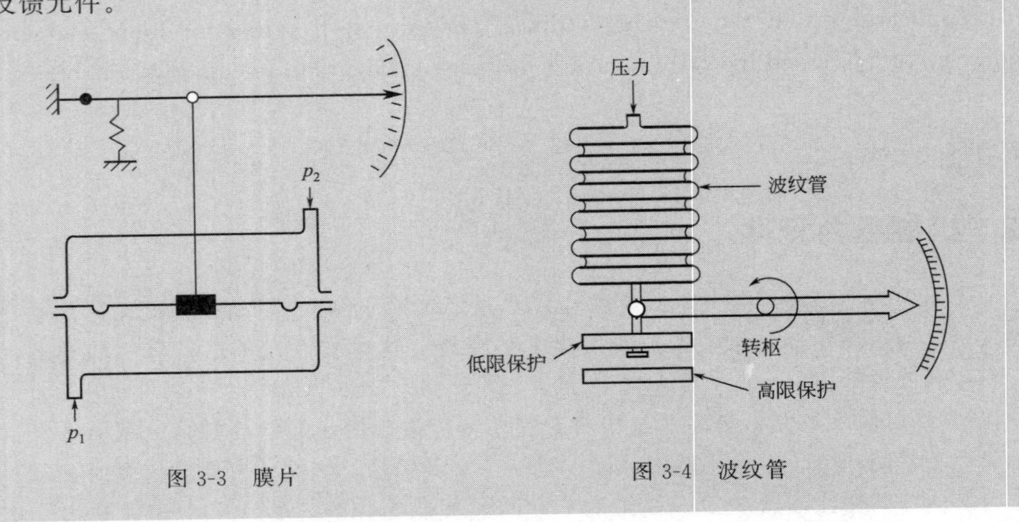

图 3-3　膜片　　　　图 3-4　波纹管

3.3 Differential Capacitance Sensors

The basic operating principle involved in differential capacitance sensors is the measurement of change in capacitance resulting from the movement of a moving plate.

$$C = \varepsilon \frac{A}{d} \tag{3-1}$$

Where

C = Capacitance between two conductors

A = the area of the plates

d = the distance of the plates

ε = the permittivity of the insulator

In this design, the sensing diaphragm is a taut metal diaphragm located equidistant between two stationary metal surfaces, comprising three plates for a complementary pair of capacitors. An electrically insulating fill fluid (usually a liquid silicone compound) transfers motion from the isolating diaphragms to the sensing diaphragm, and also doubles as an effective dielectric for the two capacitors.

In this differential capacitance sensor (see Figure 3-5), the pressure activates sensing diaphragm (moving plate) that is mounted between two stationary plates, this causes a capacitance change, which is measured in the electronic circuitry as a direct relation to pressure. Since capacitance between conductors is inversely proportional to the distance separating them, capacitance on the low-pressure side will increase while capacitance on the high-pressure side will decrease. Depending on the reference pressure used, the differential capaci-

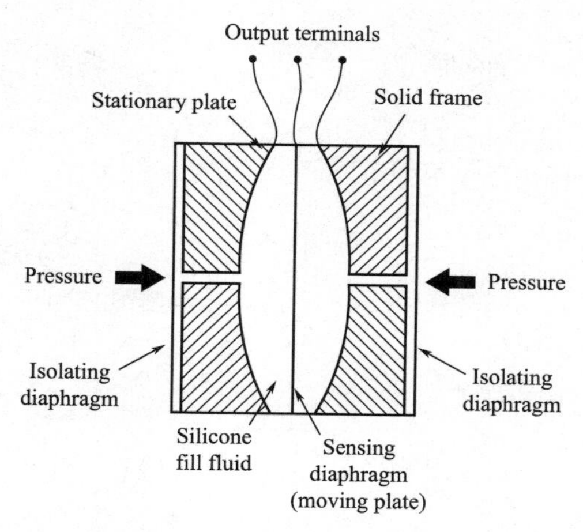

Figure 3-5 Differential capacitance sensor

tance sensor can detect absolute, gauge, or differential pressures.

A classic example of a pressure instrument based on the differential capacitance sensor is the Rosemount model 1151 differential pressure transmitter. It provides excellent response, resolution, linearity, repeatability, and stability.

In addition, since differential pressure transmitters are small and have low mass, inertia forces are low where vibration is present. However, they are relatively expensive and are sensitive to stray magnetic fields if they are not well designed. Capacitive transducers are also affected by temperature changes.

【译文】

3.3 差动电容传感器

差动电容传感器的基本工作原理是测量由于活动极板的移动而引起的电容变化。

$$C=\varepsilon\frac{A}{d} \tag{3-1}$$

式中，C 是两个导体间的电容；A 是极板面积；d 是板间距；ε 是板间介电常数。

在本设计中，传感膜片是一个张紧的金属膜片，位于两个固定的金属极板之间，距离相等，形成三个互补的电容器板。电绝缘的填充液（通常是液态有机硅化合物）将差压引起的运动从隔离膜片传递到传感膜片，并且还兼作两个电容器的有效电介质。

在这种差动电容传感器（图 3-5）中，两侧压力激活了安装在两个固定板之间的传感膜片（移动板），这会引起电容变化，该电容变化在电子电路中被测量并转换为对应的压力。由于导体之间的电容与它们之间的距离成反比，因此低压侧的电容将增加，而高压侧的电容将减小。根据参考压力，差动电容传感器可以检测绝对压力、表压或差压。

罗斯蒙特 1151 型差压变送器是典型的基于差动电容传感器的压力仪表，它具有良好的响应特性、分辨率、线性、重复性和稳定性。

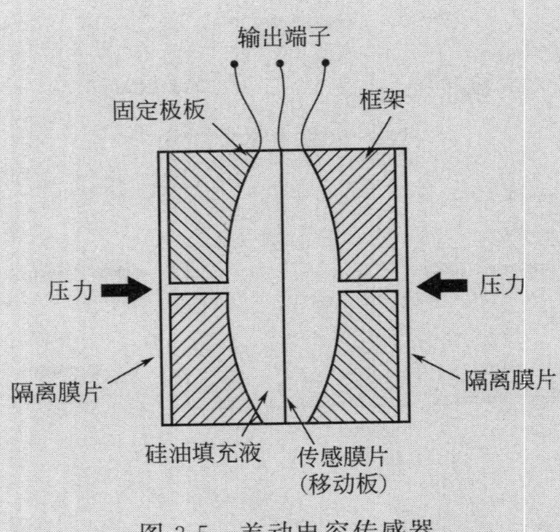

图 3-5　差动电容传感器

此外，由于它们体积和质量较小，所以在有振动的地方惯性力也小。另外，电容式传感器相对昂贵；而且如果设计不好，它们对杂散磁场很敏感；也会受到温度变化的影响。

3.4 Strain Gauge Pressure Sensors

A strain gauge is a device that changes resistance when stretched. One form of the strain gauge is a metal wire of very small diameter that is attached to the surface of a device being measured. Attaching a strain gauge to a diaphragm results in a device that changes resistance with applied pressure. Pressure forces the diaphragm to deform, which in turn causes the strain gauge to change resistance. By measuring this change in resistance, we can infer the amount of pressure applied to the diaphragm.

There are two basic types of strain gauge, bonded and unbonded, each utilizing wire or foil, but both working in the same electrical manner. A thin wire (or foil strip), usually made from chrome-nickel alloys and sometimes platinum, is subjected to stretching, and hence its resistance increases as its length increases.

One of the bonded designs is shown in Figure 3-6. Here the process pressure is applied to a flat diaphragm. The strains resulting from the diaphragm deflection are sensed by four

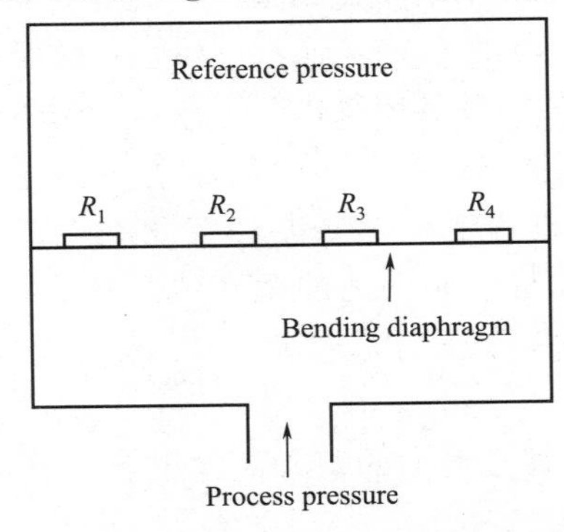

Figure 3-6　Strain gauge transducer with diaphragm element

strain elements that are bonded directly to the underside of the diaphragm. The changes are resistance of the elements that are measured as an indication of process pressure.

The working element of the strain gauge transducer shown in Figure 3-7 is a tube closed on one end, with the other end open to the process pressure. Four strain gauges are bonded to the outside of this tube. Two of the elements are strained under pressure and two are not because they are mounted longitudinally and circumferentially. When the tube is pressurized, its minute expansion changes the resistance of the gauges, which are connected to a Wheatstone bridge (see Figure 3-8). The change in output voltage of the bridge is proportional to the pressure. Calibration and terminal adjustment resistors are provided outside the tube.

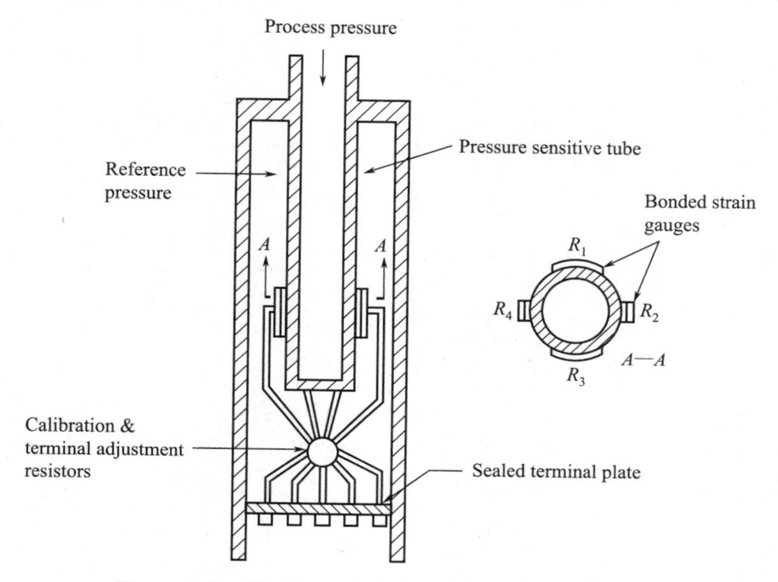

Figure 3-7　Working elements bonded to tube surface

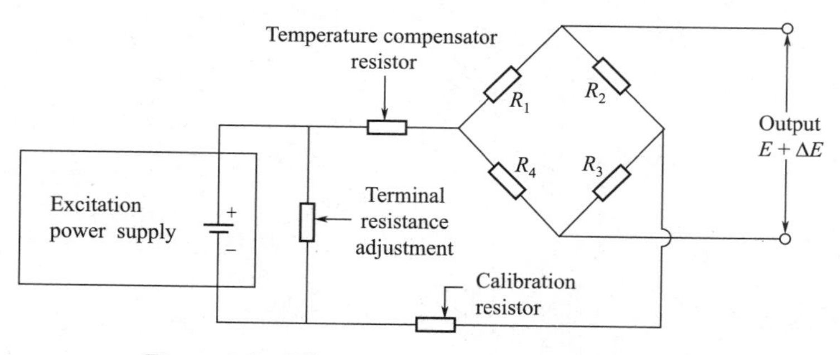

Figure 3-8　Wheatstone circuit for strain gauges

【译文】

3.4　应变片压力传感器

应变片是一种在拉伸时其电阻会改变的装置。应变片的一种形式是用直径很小的金属线连接到被测设备的表面。将应变片黏合到膜片上时，就形成了一种随着压力而改变应变片电阻的装置。压力迫使膜片变形，从而导致应变片电阻变化。通过测量电阻的变化，我们可以推断出施加在膜片上的压力。

有两种基本形式的应变片：黏合式和非黏合式，但都是利用电线或箔，以相同的电气方式工作。由铬镍合金（有时是铂）制成的细金属丝（或金属箔条）受到拉伸，其电阻值随长度的增加而增加。

黏合设计的原理如图 3-6 所示。过程压力施加到一个平坦的膜片上，膜片挠度产生的应变由四个直接黏合在膜片下方的应变元件感应到，电阻值的变化指示出过程压力的变化。

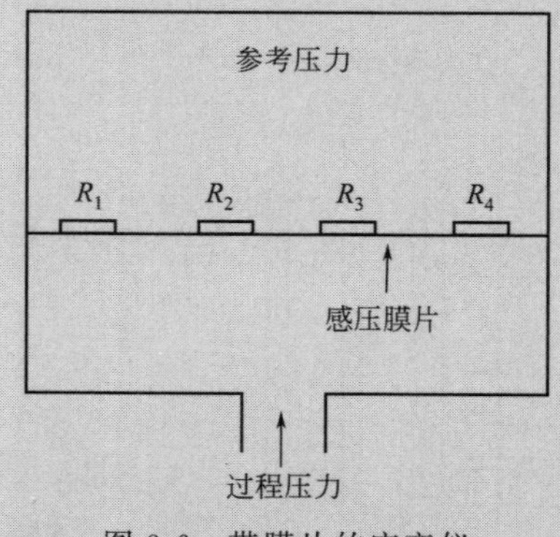

图 3-6　带膜片的应变仪

图 3-7 是一个应变仪的工作部件，它是一端封闭的管子，另一端接受过程的压力。四个应变片黏在测量管的外面，由于它们是纵向和横向安装的，所以其中两个应变片在压力作用下是变化的，另外两个则不变。当测量管受压时，其微小的膨胀变形改变了这些连接在惠斯顿电桥里的电阻（图 3-8），导致电桥的输出电压变化，并与压力成正比。校准和终端调整电阻安装在测量管外部。

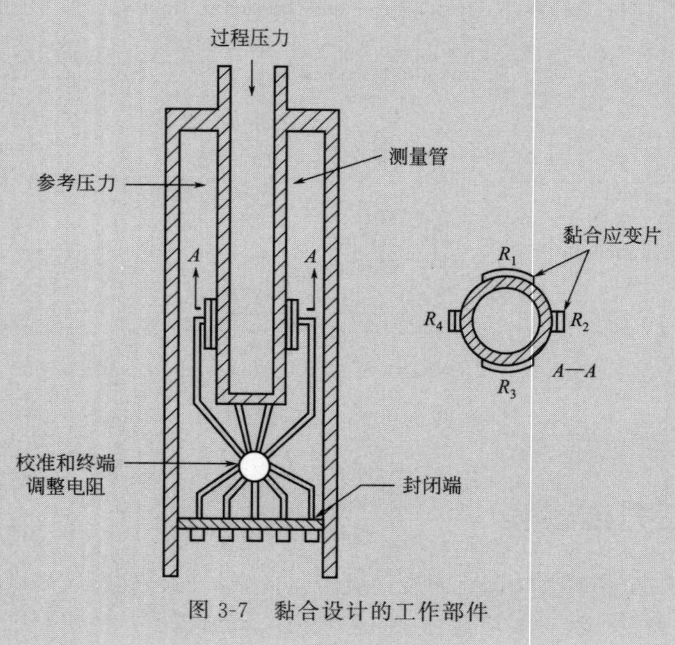

图 3-7　黏合设计的工作部件

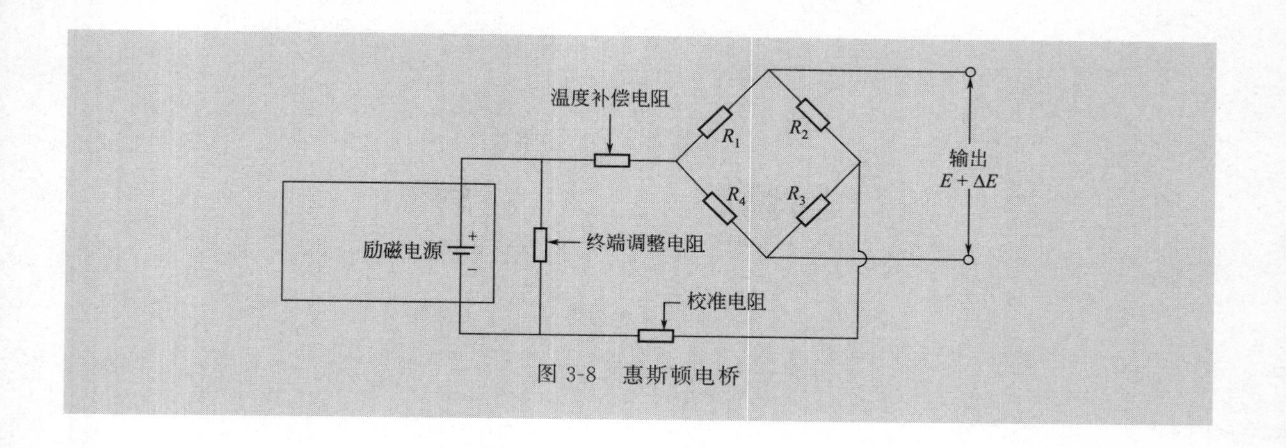

图 3-8　惠斯顿电桥

Words and Expressions　词汇和短语

1. manometer　/mə'nɑmətə/　n. 液压计
2. dialgauge　/'daiəlgeidʒ/　n. 百分表
3. absolute pressure　绝对压力
4. gage pressure　表压力
5. differential pressure　差压
6. atmospheric pressure　大气压
7. gear　/ɡɪr/　n. 齿轮
8. pivot　/'pɪvət/　n. 枢轴
9. bellows　/'belouz/　n. 波纹管
10. concentric corrugation　同心波纹
11. deflection　/dɪ'flekʃn/　n. 挠度
12. blockage　/'blɑːkɪdʒ/　n. 堵塞
13. capsule　/'kæpsuːl/　n. 膜盒
14. permittivity　/ˌpəːmɪ'tɪvəti/　n. 介电常数
15. insulator　/'ɪnsəleɪtər/　n. 绝缘体
16. silicone　/'sɪlɪkoun/　n. 硅
17. inertia force　惯性力
18. vibration　/vaɪ'breɪʃn/　n. 振动
19. stray magnetic　杂散磁场
20. strain gauge　n. 应变片
21. bonded　/'bɑndɪd/　黏合式
22. foil　/fɔɪl/　n. 箔
23. longitudinally　/lɔːŋdʒi'tjuː dinəli/　adv.　纵向地
24. circumferentially　/sɜː kʌm'fərenʃəlɪ/　adv. 周向地
25. Wheatstone bridge　惠斯顿电桥

Chapter 4
Flow Measurement

第 4 章
流量测量

4.1 Introduction

Flow can be defined as a volume of fluid in a pipe passing a given point per unit of time. This can be expressed by

$$Q=AV \tag{4-1}$$

Where A is the cross-sectional area of the pipe, and V is the average fluid velocity. Therefore, the mass flow may then be defined as

$$M=Q\rho \tag{4-2}$$

Where ρ is the density of this fluid. As temperature changes, the density of a fluid will change as well. That, in turn, may affect the accuracy of the reading unless compensation is implemented. For gases, pressure and temperature must be compensated for, if the measured values differ from the ones used for calculations. Unlike gases, liquids are incompressible but they may require temperature compensation since their density may vary significantly after a large change in temperature.

In selecting a flowmeter for any application, there are many different considerations to be made. One such consideration is the purpose the measurement is to serve. Is the measurement for accounting or custody transfer for which high accuracy and rangeability are very important? Do you require a meter for rate of flow or total flow? Is local indication or a remote signal required? If remote output is required, is it to be a proportional signal such as 4 to 20mA or is it to be a contact closure to stop or start another device such as a pump or solenoid valve or even a contact to warn of a potential hazard?

Other parameters include the physical properties of the fluid to be metered. Is it a gas or a liquid at the point of measurement? Is it a clean liquid, or is it a slurry? Is it electrically conductive? What is the operating temperature and the operating pressure?

All these parameters have a bearing on flowmeter selection. A great many types of flowmeters are available, but each type has some limitations. Flowmeters can be classified into four types:

- Volumetric, such as oval-gear flowmeters. They measure volume directly.
- Velocity, such as magnetic and turbine meters. These meters determine total flow by multiplying the velocity by the area through which the fluid flows.
- Inferential, such as differential-pressure (DP), target, and variable-area meters. These meters infer the flow by some other physical property such as differential pressure and then experimentally correlate it to flow.
- Mass, such as coriolis mass flowmeters. These devices measure mass directly.

【译文】

4.1 简介

流量可以定义为单位时间内通过管道中某个截面的流体体积，它表示为：

$$Q=AV \tag{4-1}$$

式中，A 为管道截面积；V 为平均流速。因此，质量流量可以定义为

$$M=Q\rho \tag{4-2}$$

式中，ρ 是流体的密度。随着温度的变化，流体的密度也会发生变化，这将会影响流量读数的准确性，因此需要进行补偿。对于气体，如果测量值与标定值不同，则必须进行压力和温度补偿。液体与气体不同，液体是不可压缩的，但它们需要进行温度补偿，因为液体的密度在温度大幅度变化后，可能会发生显著变化。

在选择流量计时，我们需要考虑许多因素，其中一个因素是测量的目的：流量计用于计量还是存储运输？需要很高的准确性和较大的量程吗？流量计用来测量瞬时流量还是总流量？流量计需要本地指示还是远程输出？如果需要远程输出，它是一个流量对应的信号（如 4～20mA），还是一个用于停止或启动某设备，如泵或电磁阀，甚至是一个危险触点的信号？

其他因素包括需要知道被测流体的物理性质，在测量状态下的流体是气体还是液体？它是一种干净的液体还是一种浆液？导电性如何？工作温度和工作压力是多少？

上述这些因素对流量计的选择都有一定的影响。市面上有很多种流量计，但每种流量计都有一定的局限性。流量计可以分为以下四种类型。

- 容积式，如椭圆齿轮流量计等，它们直接测量体积。
- 速度式，如电磁流量计、涡轮流量计等，这些仪表通过将速度乘以流体流过的面积来确定总的流量。
- 间接式，如差压流量计、靶式流量计和转子流量计等，这些仪表通过其他的物理特性如差压变化来推断出流量。
- 质量式，如科里奥利质量流量计等，这些仪表直接测量流体的质量。

4.2 Differential-Pressure (DP) Flowmeters

The detection of pressure drop across a restriction is the most widely used method of industrial flow measurement. The pressure decrease that results from a flowing stream passing through a restriction is proportional to the flow rate and to fluid density. Therefore, if the density is constant, the pressure drop can be interpreted into a reading of flow. This relationship is described by the following formula:

$$Q=K\,(\text{constant})\times\sqrt{\frac{\Delta p}{\rho}} \tag{4-3}$$

The most common types of differential-pressure (DP) flowmeters are the orifice plates, venturi tubes and flow nozzles.

A standard orifice plate (see Figure 4-1) is simply a smooth disk with a round, sharp-edged inflow aperture and mounting rings. In the case of viscous liquids, the upstream edge of the bore can be rounded. The shape of the opening and its location do vary widely, and this is dependent on the material being measured. Most common are concentric orifice plates with a round opening in the center.

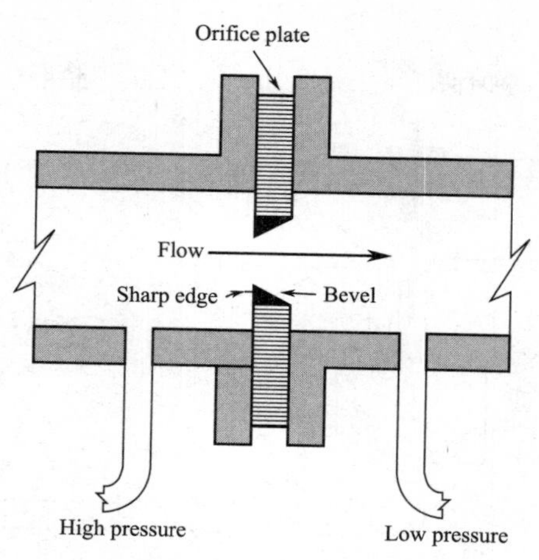

Figure 4-1 Orifice plate

The shapes of venturi tubes and flow nozzles (see Figure 4-2) have been obtained with the goal of minimizing the pressure drop across them. These tubes are often installed to reduce the size of (and therefore capital expenditures on) pumping equipment and to save on pumping energy costs. In contrast with the sharp-edged orifice, these tubes and nozzles are resistant to abrasion and can also be used to measure the flow of dirty fluids and slurries. They are, however, considerably larger, heavier, and more expensive than the orifice plate. Their installation is also more difficult.

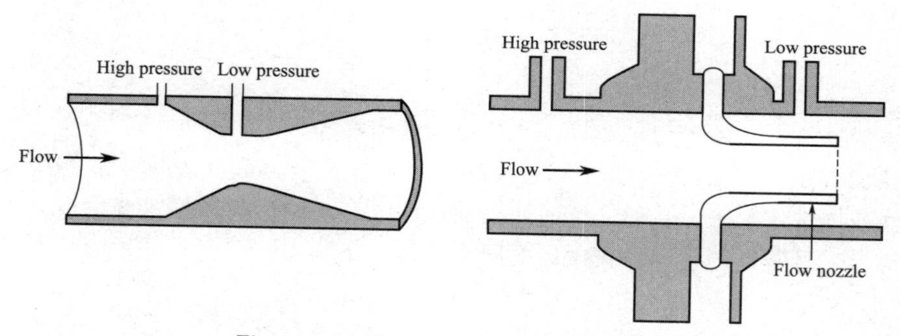

Figure 4-2 Venturi tube and flow nozzle

A typical flow measurement loop can be represented by the installation shown in Figure 4-3. The high and low-pressure taps of the primary device (orifice type shown) are fed by sensing lines to a differential-pressure flow transmitter (FT). The square root extractor is an electronic (or pneumatic) device that takes the square root of the signal from the flow transmitter and outputs a corresponding linear flow signal. This system would produce a 4 to 20mA signal that is linear with the flow rate.

Differential-pressure transmitters are typically equipped with three-valve manifolds, which are sometimes integral to the transmitters. The integral manifolds are of unitized construction and when compared to part-assembled units, they provide fewer leak points, reduced material and labor costs (especially when supplied with the transmitter), and require less physical space. On toxic and hazardous fluids, a five-valve manifold with drain or vent legs to a safe location is frequently provided, and the impulse lines are flanged or welded, instead of threaded.

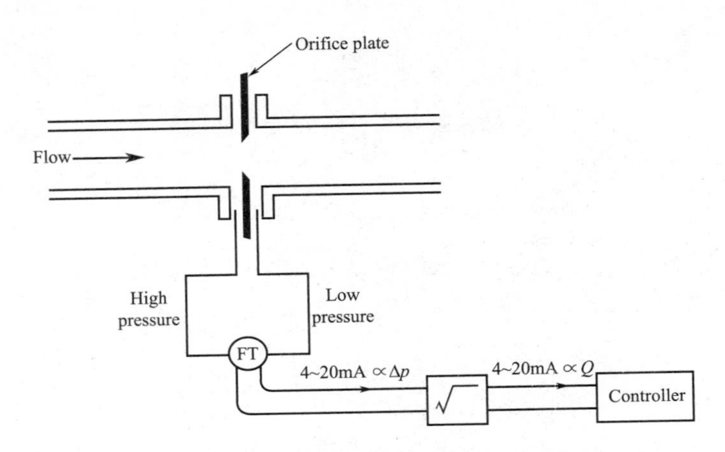

Figure 4-3　A typical flow measurement loop

【译文】

4.2　差压流量计

通过测量节流装置前后的压降是流量测量中应用最广泛的一种方法。流动的流体通过节流装置所引起的压力下降与流量和密度成比例，因此，如果密度恒定，则压降就可以用来表示流量。这种关系可以用下面的公式来描述：

$$Q = K(\text{常量}) \times \sqrt{\frac{\Delta p}{\rho}} \tag{4-3}$$

差压流量计最常见的类型是孔板、文丘里管和喷嘴。

标准孔板（图 4-1）只是一个光滑的圆盘，上面有一个圆形且边缘光滑的流入孔和安装环。在黏性液体的情况下，孔的上游边缘是圆形的，开口的形状和位置变化很大，这取决于被测量的材料。最常见的是圆孔在中心的同心孔板。

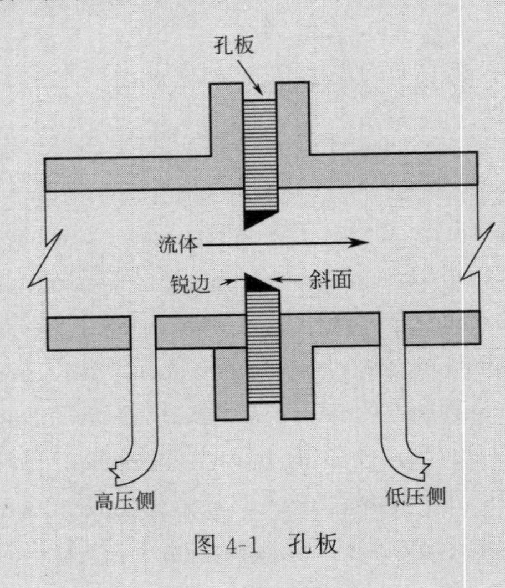

图 4-1　孔板

文丘里管和喷嘴（图 4-2）的设计形状，主要是为了使通过它们的压降最小化。在管道中安装这些设备，通常是为了减小泵设备的尺寸（减少泵设备的投资）和节省泵设备的能源成本。与锋利的孔板相比，文丘里管和喷嘴更耐磨损，也可用于测量脏污液体和泥浆的流量。但它们比孔板更大、更重，也更贵，安装也更加困难。

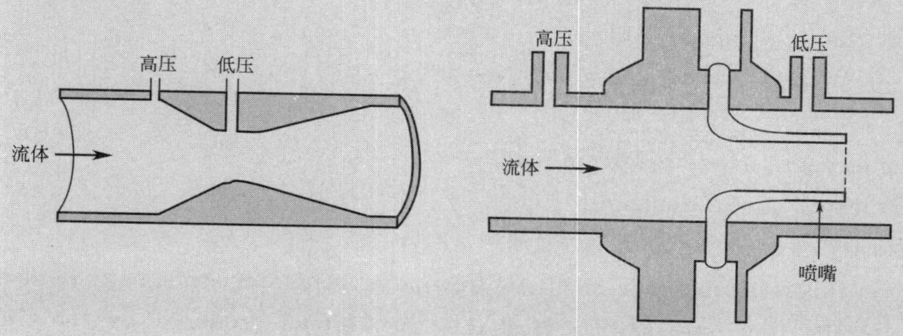

图 4-2　文丘里管和喷嘴

图 4-3 是一个典型的流量测量回路的安装形式。一次测量元件（图 4-3 所示的是孔板）两侧的高压和低压，由引压管线送入差压流量变送器（FT）。开方器是一种电子（或气动）装置，它从流量变送器中提取差压信号的平方根，并输出相应的线性流量信号。该系统将产生与流量呈线性关系的 4～20mA 的信号。

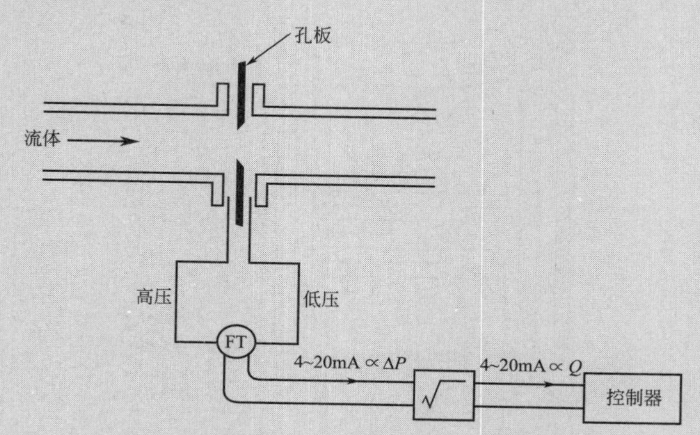

图 4-3　典型流量测量回路的安装形式

差压变送器通常配有与变送器一体的三阀组，采用单元化结构。与部分组装的部件相比，一体化差压变送器不容易泄漏，减少了材料和人工成本，并且体积更小。对有毒和危险的流量进行测量时，通常还会提供一个带有排液阀或排气阀的五阀组，并将引压管线通过法兰连接或焊接，而不采用螺纹连接。

4.3 Magnetic Flowmeters

Most industrial liquids can be measured by magnetic flowmeters, these include acids, bases, water and aqueous solutions. However some exceptions are most organic chemicals and refinery products which have insufficient conductivity for measurement. Also pure substances, hydrocarbons and gases cannot be measured.

The magnetic flowmeter's design is based on Faraday's law of magnetic induction. Faraday's law states that the voltage induced across a conductor as it moves at right angles through a magnetic field is proportional to the velocity of that conductor. This relationship is described by the following formula:

$$e = BLv \tag{4-4}$$

where

B = the strength of the magnetic field

L = the length of the conductor

v = velocity of the conductor

Based on this principle, the magnetic flowmeter generates a magnetic field that is perpendicular to the flow stream and measures the voltage produced by the fluid passing through the meter as detected by the electrodes (see Figure 4-4). The voltage produced by the magnetic flowmeter is proportional to the average velocity of the volumetric flow rate of the conductive fluid.

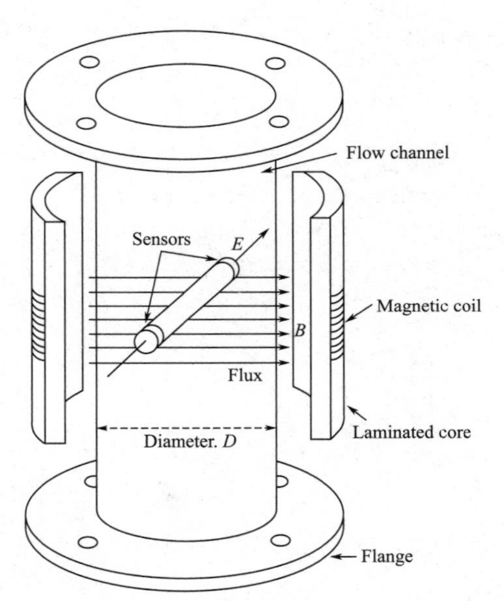

Figure 4-4　Magnetic flowmeter

The relationship between the volume of liquid Q and the velocity may be expressed as:

$$Q = Av$$

Writing the area, A (m^2), of the pipe as

$$A = \frac{\pi D^2}{4}$$

gives the induced voltage as a function of the flow rate, that is,

$$e = BDv = \frac{BDQ}{A} = \frac{4BQ}{\pi D} \tag{4-5}$$

If we wish to have a formula defining flow rate Q in terms of voltage, we may simply manipulate the last equation to solve for Q:

$$Q=\frac{\pi D}{4B}e \tag{4-6}$$

The magnetic flowmeter's tube is constructed of non-magnetic material (to allow magnetic field penetration) such as stainless steel and is lined with a suitable material to prevent short-circuiting the voltage generated between the electrodes. The tube is also used to support the coils and transmitter assembly. Generally, the electrodes are made of stainless steel, but other materials are also available (choose with care to avoid corrosion). Dirty liquids may foul the electrodes, and cleaning methods such as ultrasonic may be required.

【译文】

4.3 电磁流量计

大多数工业流体可以用电磁流量计进行测量，这些物质包括酸、碱、水和水溶液。但也有一些例外，大多数有机化学品和炼油产品的导电性不足，故无法用电磁流量计进行测量；纯物质、碳氢化合物和气体也无法测量。

电磁流量计的设计原理是基于法拉第磁感应定律的。法拉第定律指出，当导体在磁场中垂直切割磁力线时，导体上感应的电压与导体的速度成正比。这种关系可以用下面的公式来描述：

$$e=BLv \tag{4-4}$$

式中，B 是磁场强度；L 是导体的长度；v 是导体的速度。

基于这一原理，电磁流量计产生一个垂直于流向的磁场，并通过电极来检测流体流过流量计的磁场所产生的电压（图 4-4）。电磁流量计产生的电压与导电流体的体积流量的平均速度成正比。

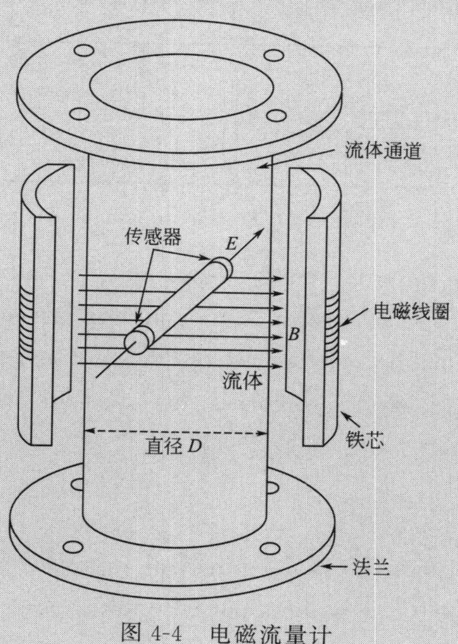

图 4-4　电磁流量计

液体体积 Q 与速度 v 之间的关系可以表示为

$$Q=Av$$

管道的面积 A（m^2）为

$$A=\frac{\pi D^2}{4}$$

可以得到感应电压和流量的函数关系为

$$e=BDv=\frac{BDQ}{A}=\frac{4BQ}{\pi D} \tag{4-5}$$

如果用感应电压直接定义流量 Q，我们可以简单地由式（4-5）求解：

$$Q=\frac{\pi D}{4B}\times e \tag{4-6}$$

电磁流量计测量管由非磁性材料（允许磁场穿透）如不锈钢制成，并内衬适当的材料以防止电极之间产生的电压短路。该测量管也用于支撑线圈和变送器组件。通常，电极由不锈钢制成，也有用其他材料的，必须认真选择以避免腐蚀。不干净的液体会污染电极，可能需要使用超声波等方法进行清洗。

4.4 Turbine Flowmeters

A turbine flowmeter (see Figure 4-5) consists of a rotor (similar to a propeller) that has a diameter almost equal to the pipe's internal diameter, which is supported by two bearings to allow the rotor to rotate freely, the rotational velocity of the rotor is proportional to the fluid velocity.

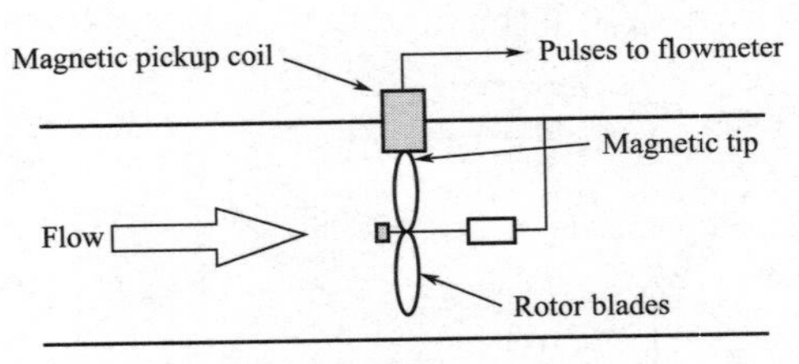

Figure 4-5 Turbine flowmeter

A magnetic pickup coil, mounted on the pipe, detects the passing of the rotor blades, generating a frequency output. In an electronic turbine flowmeter, volumetric flow is directly and linearly proportional to pickup coil output frequency. Each pulse represents the passage of a calibrated amount of fluid. We may express this relationship in the form of an equation:

$$Q=\frac{f}{k} \tag{4-7}$$

where

f = Frequency of output signal (Hz, equivalent to pulses per second)

Q = Volumetric flow rate (e. g. gallons per second)

k = "K" factor of the turbine element (e. g. pulses per gallon)

Turbine meters require a good laminar flow. In fact 10 pipe diameters of straight line up stream, and no less than 5 pipe diameters downstream from the meter are required. They are therefore not accurate with swirling flows.

The turbine meter is easy to install and maintain. It is bidirectional, has a fast response, and is compact and lightweight. The device is not sensitive to changes in fluid density (though at very low specific gravities, rangeability may be affected), and it can generate a pulse output signal to directly operate digital meters.

However, turbine meters do have disadvantages. They are not recommended for measuring steam since condensate does lubricate the bearings well, though some designs will handle steam measurement. Also, they are sensitive to dirt and cannot be used for highly viscous fluids or for fluids with varying viscosity. Flashing, slugs of vapor, or gas in the liquid produce blade wear and excessive bearing friction, which results in poor performance and possible turbine damage. In addition, turbine meters are affected by air and gas entrained in the liquid (in amounts exceeding 2 percent by volume; therefore, the pipe must be full). Strainers may be required up stream to minimize particle contamination of the bearings.

【译文】

4.4 涡轮流量计

涡轮流量计（图 4-5）内部安装有一个叶轮（类似于一个螺旋桨），直径约等于管道内部直径，由两个轴承支撑其自由旋转。叶轮的旋转速度与流体流速成比例。

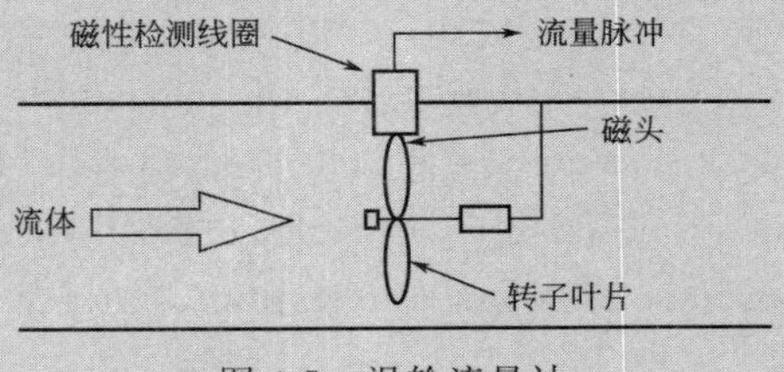

图 4-5 涡轮流量计

安装在管道上的电磁式拾波器检测通过的叶片，产生频率输出。在电子涡轮流量计中，体积流量与拾波器的输出频率成正比。每个脉冲代表一个校准量的流体通过。我们可以用方程的形式来表示这种关系：

$$Q=\frac{f}{k} \tag{4-7}$$

式中，f 是输出信号的频率，Hz；Q 是体积流量，gal[1]/s；k 是涡轮元件因数，1/gal。

涡轮流量计需要良好的层流。通常入口直管段的长度是管道内径的 10 倍，出口直管段的长度不小于管道内径的 5 倍。涡轮流量计对于旋流的测量是不准确的。

[1] $1gal/s=3.785\times10^{-3}m^3/s$。

涡轮流量计易于安装和维护。它是双向的，响应速度快，结构紧凑，重量轻。该装置对流体密度的变化不敏感（虽然在很低的密度下，量程可能会受到影响），它可以产生脉冲输出信号直接驱动数字仪表。

涡轮流量计也有缺点，一般不用于测量蒸汽，因为凝结水会影响轴承，尽管有些会设计蒸汽处理装置。此外，它们对污垢很敏感，不能用于高黏度流体或具有可变黏度的流体。液体中的水蒸气或气体的闪蒸、气团会造成叶片的磨损和轴承的过度摩擦，从而导致性能低下甚至可能损坏涡轮。此外，涡轮流量计还受到液体中夹带的空气和气体的影响（按体积计算超过 2%，因此管道必须是满的）。此外，可能需要在上游安装过滤器，以减少颗粒对轴承的磨损。

4.5 Coriolis Mass Flowmeters

Volumetric flowmeters are subject to ambient and process changes, such as density, that change with temperature and pressure. Sometimes the meter user may be more interested in the weight (mass) of the gas or liquid.

Mass flow measurement can be categorized as true mass flow measurement or inferential mass flow measurement. In the former, the mass flow is measured directly without regard to separately measured density or other physical properties. For inferential measurement, it is necessary to measure properties of the process such as temperature and pressure of a gas to infer the mass flow. For fluids with unknown properties, true mass flow measurement will be required. For example, we can use Coriolis flowmeters.

The first industrial Coriolis mass flowmeter was introduced in 1972. At that time a Coriolis mass flowmeter was very bulky, not very accurate, not reliable, and very expensive. Today, many manufacturers offer a large number of different designs, from a single straight tube, to dual straight tubes, to all types of bent single tube or dual tubes. The latest generation of Coriolis mass flowmeters is accurate and reliable, and some of the designs are no longer bulky or application sensitive. Coriolis mass flowmeters have a very large installation base in liquid, slurry, and gas applications, and have been installed in some steam applications.

In the Coriolis effect design (see Figure 4-6), one tube is forced to oscillate at its natural frequencies perpendicular to the flow direction, as mass flow rate through the tube increases, so does the degree of twisting. By monitoring the severity of this twisting motion, we may infer the mass flow rate of the fluid passing through the tube. In order to reduce the amount of vibration generated by a Coriolis flowmeter, and more importantly to reduce the effect any external vibrations may have on the flowmeter, two identical tubes are built next to each other and shaken in complementary fashion (always moving in opposite directions), this eliminates the effect of any common-mode vibrations on the inferred flow measurement.

A problem unique to Coriolis flowmeters is the entrapment of gas bubbles (in a liquid process) or liquid droplets (in a gas process). Either condition will create an uneven distribution of mass inside the flowmeter's tubes. The tubes of a Coriolis flowmeter should be ori-

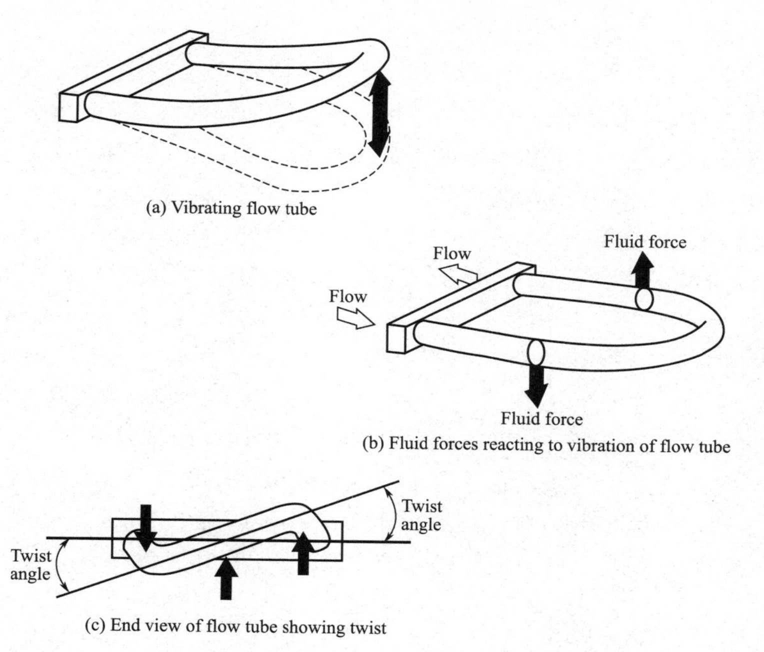

Figure 4-6 Coriolis effect design

ented such that bubbles or droplets cannot collect within them (see Figure 4-7). For liquid processes, the tubes should be located below the pipe's centerline. For gas processes, the tubes should be located above the pipe's centerline.

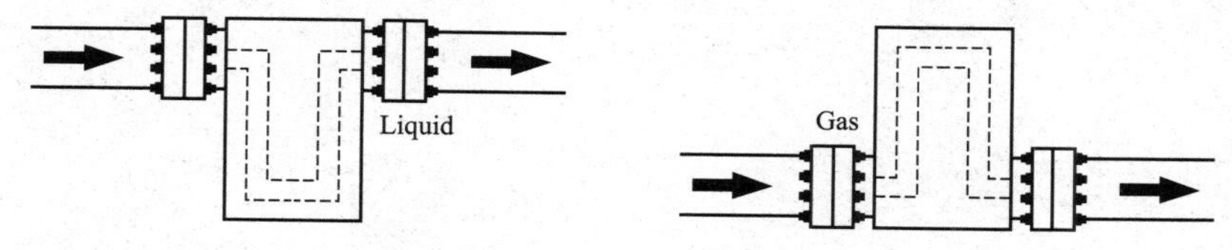

Figure 4-7 Proper installation

The Coriolis flowmeter has many advantages. It directly measures mass flow and density, and some also measure temperature. It handles difficult applications, is applicable to most fluids, has no Reynolds number limitation, and is not affected by minor changes in specific gravity and viscosity. In addition, the Coriolis flowmeter device requires low maintenance. On the other hand, the purchase cost of Coriolis flowmeters is high, and inaccuracies are introduced from air and gas pockets in the liquid as well as by slug flow. The pipe must be full and must remain full to avoid trapping air or gases inside the tube. A high-pressure loss is generated due to the small tube diameters.

【译文】

4.5 科里奥利质量流量计

容积式流量计受环境和过程变化的影响，如密度随温度和压力的变化而变化。在某些场合，仪表使用者对气体或液体的重量（质量）更感兴趣。

质量流量的测量可分为直接式质量流量检测和间接式质量流量检测。前者质量流量是直接测量的，不需要考虑单独测量密度或其他物理性质。而间接式质量流量检测，必须测量过程的特性，如温度和压力，以推断出质量流量。对于性质未知的流体，需要进行直接的质量流量测量，例如可以使用科里奥利流量计。

第一台工业科里奥利质量流量计于1972年问世。当时的科里奥利质量流量计体积庞大，精度不高，可靠性差，价格昂贵。今天，许多制造商提供了大量不同的设计，从单直管到双直管，到所有类型的弯曲单管或双管。新一代科里奥利质量流量计准确可靠，体积小而精度高。科里奥利质量流量计在液体、泥浆和气体的流量测量中有大量的应用，并且已经应用于蒸汽测量。

在科里奥利效应设计（图4-6）中，一根测量管被迫以其垂直于流动方向的固有频率振荡，当通过测量管的质量流量增加时，扭转的程度也增加。通过检测扭转运动的严重程度，可以推断出流体通过管道的质量流量。为了减少科里奥利流量计产生的振动量，更重要的是为了减少外部振动对流量计的影响，双测量管方式采用了两个相同的测量管彼此相邻，并以互补的方式摇动（始终朝相反的方向移动），从而消除了共模振动对流量测量的影响。

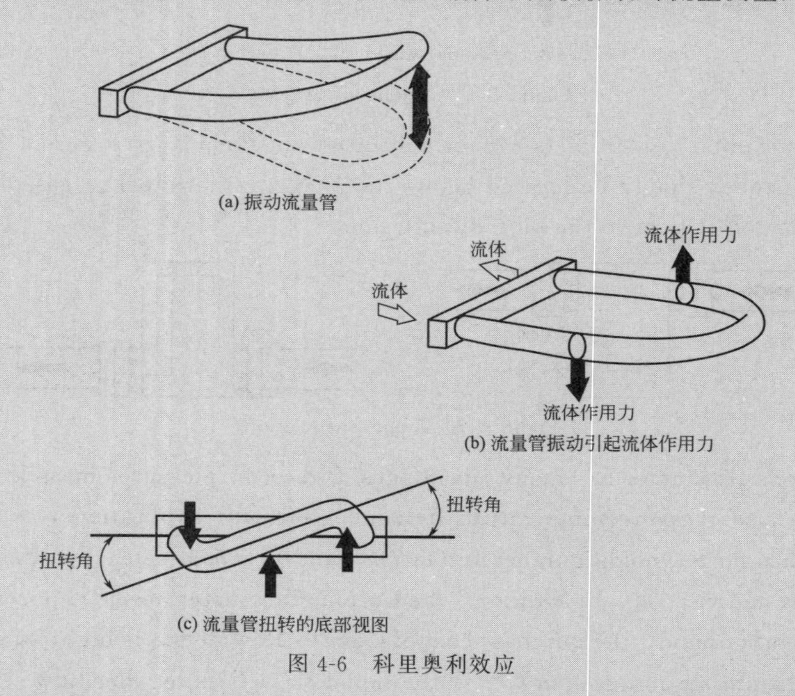

图4-6 科里奥利效应

科里奥利流量计特有的一个问题是气泡（在液体过程中）或液滴（在气体过程中）的夹持。这两种情况都会造成流量计管内质量分布不均匀。科里奥利流量计的测量管应定向安装，使气泡或液滴不能在测量管内聚集（图4-7）。测量液体时，测量管应该位于管道的中心线以下；测量气体时，测量管应该位于管道的中心线以上。

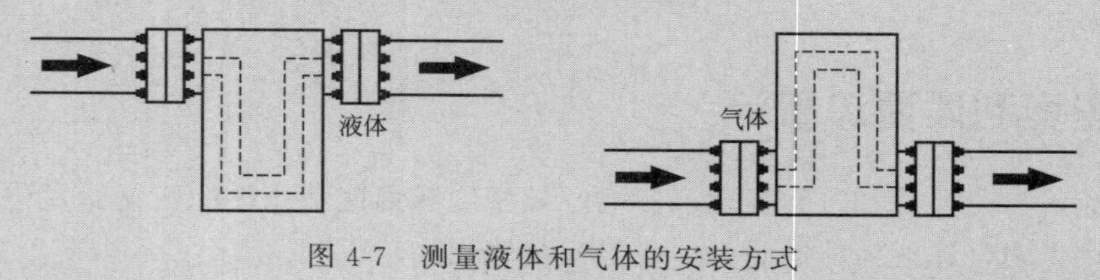

图4-7 测量液体和气体的安装方式

科里奥利流量计具有许多优点。它可以直接测量质量流量和密度，有些还可测量温度。它能处理困难的应用，适用于大多数流体，没有雷诺数限制，不受密度和黏度微小变化的影响，易于维护。另一方面，科里奥利流量计的购置成本高，存在气穴、段塞流等现象。流量管道必须始终是满的，以避免将空气或气体困在管道内。由于管径小，会产生压力损失。

Words and Expressions 词汇和短语

1. density /ˈdensəti/ n. 密度
2. compensation /ˌkɑːmpenˈseɪʃn/ n. 补偿
3. hazard /ˈhæzərd/ n. 危险
4. conductive /kənˈdʌktɪv/ adj. 导电的
5. volumetric /ˌvɑljəˈmɛtrɪk/ adj. 体积的
6. oval-gear flowmeter 椭圆齿轮流量计
7. turbine meter 涡轮流量计
8. Coriolis mass flowmeter 科里奥利质量流量计
9. pressure drop 压降
10. orifice plate 孔板
11. venturi tube 文丘里管
12. flow nozzle 喷嘴
13. viscous /ˈvɪskəs/ adj. 黏性的
14. abrasion /əˈbreɪʒn/ n. 磨损
15. square root extractor 开方器
16. three-valve manifold 三阀组
17. impulse line 引压管线
18. flange /flændʒ/ n. 法兰
19. thread /θred/ n. 螺纹
20. magnetic flowmeter 电磁流量计
21. aqueous /ˈeɪkwiəs/ adj. 含水的
22. hydrocarbon /ˌhaɪdrəˈkɑː rbən/ n. 碳氢化合物
23. penetration /ˌpenəˈtreɪʃn/ n. 渗透
24. stainless steel 不锈钢
25. short-circuiting 短路
26. electrode /ɪˈlektroʊd/ n. 电极
27. rotor /ˈroʊtər/ n. 转子
28. turbine flowmeter 涡轮流量计
29. laminar flow 层流
30. swirling flow 旋涡流
31. lubricate /ˈluːbrɪkeɪt/ vt. 使……润滑
32. viscosity /vɪˈskɑːsəti/ n. 黏性
33. strainer /ˈstreɪnər/ n. 过滤器
34. oscillate /ˈɑːsɪleɪt/ vt. 使振荡
35. twist /twɪst/ v. 使弯曲
36. severity /sɪˈverəti/ n. 严重性
37. Reynolds number 雷诺数

Chapter 5

Level Measurement

第5章

液位测量

5.1 Introduction

Level measurement is defined as the measurement of the position of an interface between two media. These media are typically gas and liquid, but they also could be two liquids. Level measurement is a key parameter that is used for reading process values, for accounting needs, and for control. By measuring the level, the operator can know how much is stored in the container. Monitor and control the level to keep it within a certain range, or to maintain a balance between the input and output of the container.

With the development of electronic technology and automation technology, the level measurement technology has greatly improved. The production of petroleum, chemical, pharmaceutical and other industries is inseparable from the monitoring of liquid level. The importance of level sensors is becoming increasingly prominent, which directly affects the safety of production and product quality. Of the typical flow, level, temperature, and pressure measurements, flow tends to be the most difficult, but level follows closely behind.

There are numerous ways to measure level that require differing technologies and to encompass all the various units of measurement. There are several common level measurement methods:

- Ultrasonic
- Pulse echo
- Pulse radar
- Pressure, hydrostatic
- Weight, strain gauge
- Conductivity
- Capacitive.

【译文】

5.1 简介

液位测量被定义为两种介质之间交界位置的测量，这些介质通常是气体和液体，但也可能是两种液体。液位测量是了解过程变化的一个关键参数，用于计量需求和控制。通过测量液位，操作人员可以获知容器中存储量的多少。监测和控制液位，可以使其保持在某一个范围以内，或者保持容器的输入与输出之间的平衡。

随着电子技术和自动化技术的不断发展，液位测量技术有了很大的提高。石油、化工、医药等行业的生产，都离不开液位的监控，液位传感器的重要性日趋凸显，直接影响着生产的安全和产品的质量。在典型的流量、液位、温度和压力测量中，液位检测的难度仅次于流量检测。

有许多方法来测量液位，需要不同的技术，并包含不同的测量单位。常见的液位测量方法有以下几种：超声波、脉冲回波、脉冲雷达、压力-流体静力学、重力-应变片式、电导式、电容式。

5.2 Differential-Pressure Level Measurement

Differential-pressure level measurement, also known as "hydrostatic," is based on the height of the liquid head (the U-tube principle).

Level measurement in open tanks is based on the formula that the pressure head is equal to the liquid height above the tap multiplied by the specific gravity of the fluid being measured. This can be expressed by

$$p = \rho g H \tag{5-1}$$

where ρ is the density of the liquid, H is the height of the liquid level.

In closed tanks, the true level is equal to the pressure measured at the tank bottom minus the static pressure above the liquid surface. In order to obtain the differential pressure, a leg is connected from the tank top to the low-pressure side of the differential pressure transmitter. Two options are available: dry leg (see Figure 5-1) and wet leg (see Figure 5-2).

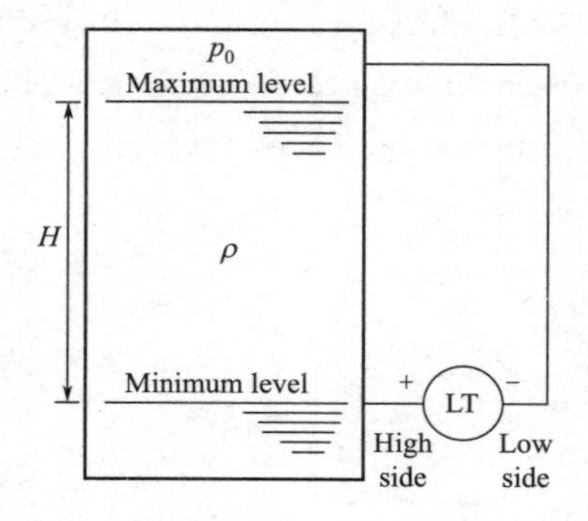

Figure 5-1 Dry leg

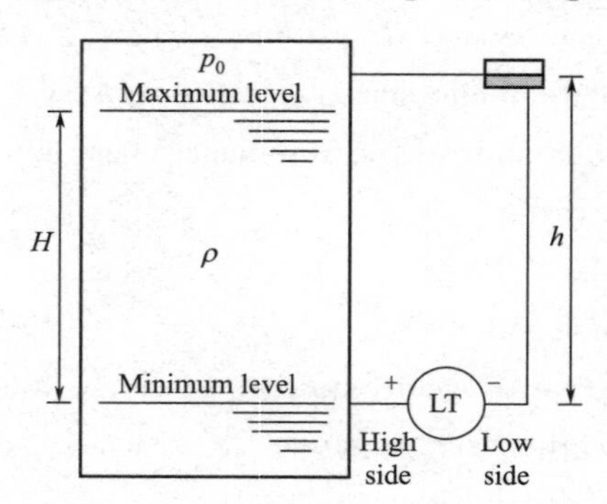

Figure 5-2 Wet leg

When the low-pressure impulse line is connected directly to the gas phase above the liquid level, it is called a dry leg. If the gas phase is condensable, say steam, condensate will form in the low-pressure impulse line resulting in a column of liquid, which exerts extra pressure on the low-pressure side of the transmitter. A technique to solve this problem is to add a knockout pot below the transmitter in the low-pressure side. Periodic draining of the condensate in the knockout pot will ensure that the impulse line is free of liquid. In dry leg applications, it is expected that the low side will remain empty. The differential-pressure can be expressed by

$$\Delta p = p_+ - p_- = \rho g H \tag{5-2}$$

In a wet leg system, the low-pressure impulse line is completely filled with liquid (usually the same liquid as the process) and hence the name wet leg. A level transmitter is used in an identical manner to the dry leg system. At the top of the low pressure impulse line is a small catch tank, the gas phase or vapour will condense in the wet leg and the catch tank. This pressure, being a constant, can easily be compensated for by calibration. The differential-pressure can be expressed by

$$\Delta p = p_+ - p_- = \rho g (H - h) \tag{5-3}$$

In some cases, it is not possible to mount the level transmitter right at the base level of the tank. Say for maintenance purposes, the level transmitter has to be mounted h meters below the base of the tank as shown in Figure 5-3.

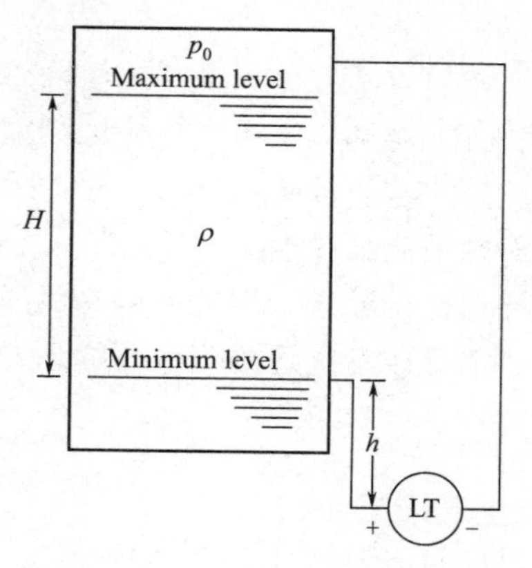

Figure 5-3 Zero suppression

In this situation, the liquid in the tank exerts a varying pressure that is proportional to its level H on the high-pressure side of the transmitter, and the liquid in the high-pressure impulse line also exerts an additional pressure on the high-pressure side, however, this additional pressure is a constant (ρgh) and is present at all times. That is, the pressure on the high-pressure side is always higher than the actual pressure exerted by the liquid column in the tank. This constant pressure would cause an output signal that is higher than 4mA when the tank is empty and above 20mA when it is full. The transmitter has to be negatively biased by a value of $-\rho gh$ so that the output of the transmitter is proportional to the tank level only. This procedure is called Zero Suppression and it can be done during calibration of the transmitter. A zero suppression kit can be installed in the transmitter for this purpose.

In the other case, when a wet leg installation is used (see Figure 5-2), the low-pressure side of the level transmitter will always experience a higher pressure than the high-pressure side. This is due to the fact that the height of the wet leg (h) is always greater than the maximum height of the liquid column (H) inside the tank. The differential pressure Δp sensed by the transmitter is always a negative number. Δp increases from $-\rho gh$ to $\rho g(H-h)$ as the tank level rises from 0% to 100%. If the transmitter were not calibrated for this constant negative error ($-\rho gh$), the transmitter output would read low at all times. To properly calibrate the transmitter, a positive bias ($+\rho gh$) is needed to elevate the transmitter output. This positive biasing technique is called zero elevation.

Differential-pressure measuring devices are easy to install and have a wide range of measurement. With proper modifications, such as extended diaphragm seals and flange connections, these instruments will handle hard-to-measure fluids (e.g., viscous, slurries, corrosive, hot). On the other hand, differential-pressure measuring devices are affected by changes in density. They should be used only for liquids with fixed specific gravity or where errors due to varying specific gravities are acceptable or compensated for.

【译文】

5.2 差压液位测量

差压液位测量法，也称为流体静力学方法，基于U形管原理。

在开放式储罐中，液位测量是根据压力等于液体的高度乘以液体的密度这个公式，它表示为

$$p=\rho gH \tag{5-1}$$

式中，ρ 是液体的密度；H 是液位的高度。

在封闭的容器中，真实液位正比于差压，即容器底部的压力减去液体表面的静压。为了获得差压，需要从容器顶部引一条测压管线连接到差压变送器的低压侧。有两种选择：干相（图5-1）和湿相（图5-2）。

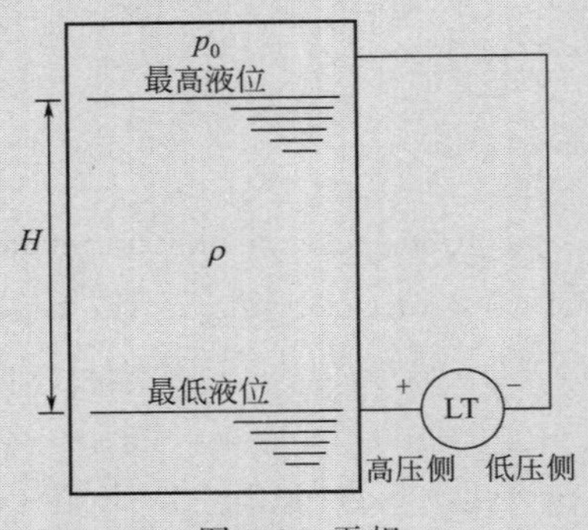

图5-1 干相

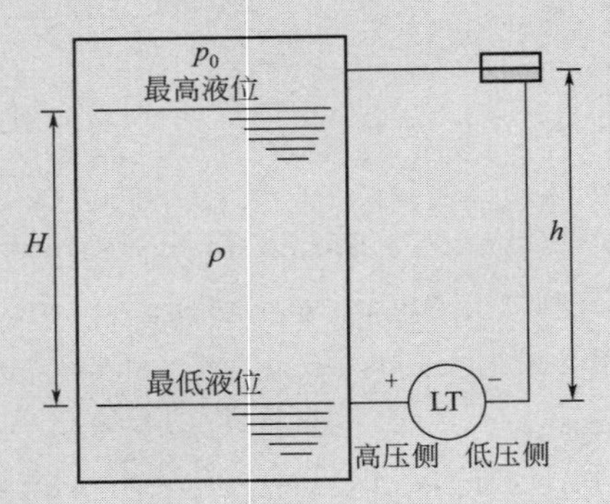

图5-2 湿相

当低压侧的引压管线直接连接到液位以上的气相时，称为干相。如果气相是可冷凝的，如蒸汽，则冷凝物会在低压侧引压管线上形成一柱液体，对变送器低压侧产生额外的压力。解决这一问题的一种方法是在变送器下方的低压侧增加一个凝液罐，抽取罐内的凝析液，定期排出，保证管线内不含液体。在干相方式的应用中，低侧将保持为空。两侧的差压表示为

$$\Delta p=p_{+}-p_{-}=\rho gH \tag{5-2}$$

在湿相系统中，低压侧的引压管线完全充满液体（通常是与工艺相同的液体），因此被称为湿相。液位变送器的使用与干相系统时一样。在低压侧的引压管线的顶部是一个小的收集槽，气相或蒸汽将在管线和集水槽中冷凝。由于这个压力是常数，很容易通过校准得到补偿。两侧的差压表示为

$$\Delta p=p_{+}-p_{-}=\rho g(H-h) \tag{5-3}$$

在某些情况下，不可能正好将液位变送器安装在容器的底部。例如，为了维护方便，有时候需要将变送器安装在容器底部以下 h 米处，如图5-3所示。

在这种情况下，容器内的液体在高压侧产生与液位 H 成正比的压力，高压侧的引压管线也产生一个额外的压力，然而这个额外压力是一个大小为 ρgh 的常数。也就是说，变送器高压侧的压力总是大于容器内液柱实际施加的压力。这个恒定的额外压力会导致当容器液位为下限时，变送器的输出信号高于4mA；而液位为上限时，输出信号超过20mA。变送器必须负偏置 $-\rho gh$ 值，使变送器的输出信号仅和容器液位成正比。这个过程被称为零点抑制，它可以在变送器的校准期间完成，为此，必须在变送器中安装一个零点抑制装置。

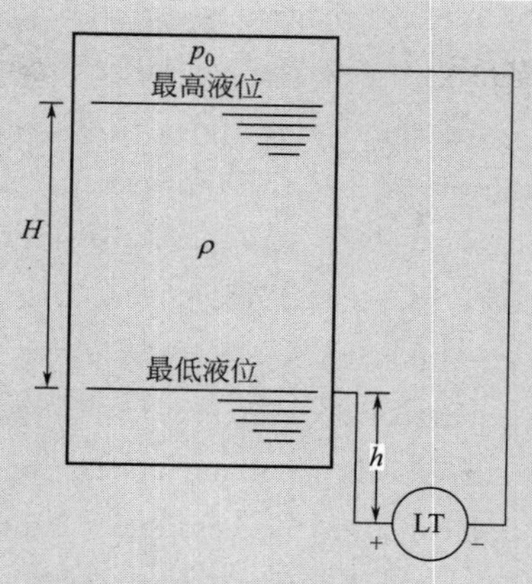

图 5-3　零点抑制

在另一种情况下，如采用湿相安装时，液位变送器低压侧的压力始终高于高压侧的压力。这是由于湿相液柱的高度（h）总是大于罐内液体柱（H）的最大高度，变送器检测到的压差 Δp 总是一个负数。当液位从 0%上升到 100%时，差压 Δp 从 $-\rho gh$ 增加到 $\rho g(H-h)$。如果变送器没有校准这个偏差，变送器的输出会偏低，所以必须正偏置 $+\rho gh$ 值，才能校准变送器的输出，这种方法称为零点抬升。

差压测量装置安装方便，测量范围广。通过适当的改进，如增加隔膜密封和法兰连接，这些仪表就能够处理难以测量的液体（例如黏性、泥浆、腐蚀性、高温液体）。另一方面，差压测量装置受密度变化的影响，它们只适用于具有固定密度的液体，或密度偏差可以接受或得到补偿的液体。

5.3 Capacitive Level Instruments

A capacitor is an electrical component capable of storing a certain electrical charge. It basically consists of two metal plates separated by an insulator known as a dielectric. The electrical size (capacitance) of a capacitor depends on

- The surface area of the plates
- The distance of the plates
- The dielectric constant of the material between the plates.

The basic principle behind capacitive level instruments is the capacitance equation:

$$C=\frac{\varepsilon A}{d} \tag{5-4}$$

Where

C = Capacitance

ε = Permittivity of dielectric (insulating) material between plates

A = Overlapping area of plates

d = Distance separating plates

When the process material replaces the empty space of air (dielectric constant 1) in the vessel, the capacitance of the capacitor increases (the higher the level, the bigger the capacitance). All liquids or solids have a dielectric constant higher than one (1).

Capacitance level instruments measure the overall capacitance of a capacitor formed by (typically, but not necessarily) the tank wall and a probe (rod/cable). As level increases, capacitance increases between the probe and the tank wall, causing the instrument to output a greater signal. This device can be used for conductive or nonconductive liquids, but the dielectric constant of the liquids being measured must remain constant.

If the liquid in the tank is conductive (see Figure 5-4), it cannot be used as the dielectric (insulating) medium of a capacitor. Consequently, capacitive level probes designed for conductive liquids are coated with plastic or some other dielectric substance, so the metal probe forms one plate of the capacitor and the conductive liquid forms the other. In this style of capacitive level probe, the variables are permittivity (ε) and distance (d), since a rising liquid level displaces low-permittivity gas and essentially acts to bring the tank wall electrically closer to the probe. This means total capacitance will be greatest when the tank is full (ε is greatest and effective distance d is at a minimum), and least when the tank is empty (ε of the gas is in effect, and over a much greater distance).

If the liquid is non-conductive (see Figure 5-5), it may be used as the dielectric itself, with the metal wall of the storage tank forming the second capacitor plate. The probe is just a bare metal cable or rod. In this style of capacitive level probe, the only variable affecting probe capacitance is permittivity (ε), provided the liquid has a substantially greater permittivity than the vapor space above the liquid. This means total capacitance will be greatest when the tank is full (average permittivity ε is at a maximum), and least when the tank is empty. Distance (d) is constant with a non-conducting process liquid, being the radius of the tank (assuming the probe is mounted in the center).

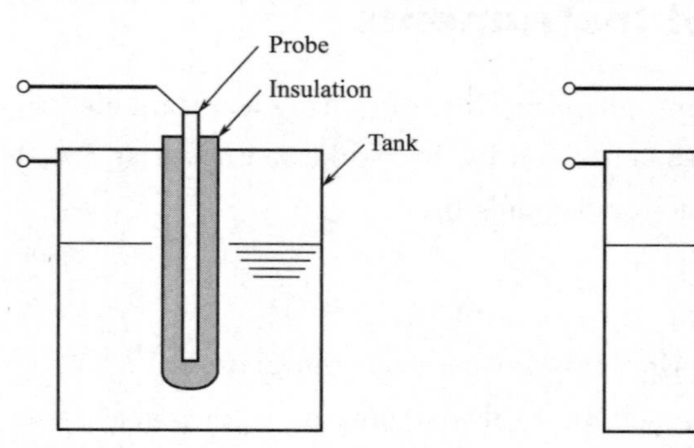

Figure 5-4 Conductive liquid   Figure 5-5 Non-conductive liquid

Capacitance level measurement is an easy technique to install. It is simply designed with no moving parts, is unaffected by nonconductive buildup, and can be used for pressurized or unpressurized vessels. However, calibration may be time consuming. The unit is affected by changes to the material's dielectric constant and thus requires temperature compensation. The installation must ensure that the probe is not in contact with the tank walls. If the application requires an insulated probe, users must take care during installation to prevent damage to the probe's insulating material.

【译文】

5.3　电容式液位计

电容器是一种能储存一定电荷的电子元件。它一般是由两块金属板组成，中间由一种叫做电介质的绝缘体隔开。电容的大小取决于极板的表面积、极板之间的距离和极板之间材料的介电常数。电容式液位计的基本原理是电容方程：

$$C=\frac{\varepsilon A}{d} \tag{5-4}$$

式中，C 是电容值；ε 是介电常数；A 是极板表面积；d 是极板之间的距离。

当电容器里的空气（介电常数为）被工艺介质取代后，电容值将增大（液位越高，电容越大），因为所有的液体或固体的介电常数都大于 1。

电容式液位计测量的是由金属罐壁和探头（棒/电缆）组成的电容器的总电容。当液位增加时，探头和罐壁之间的电容增加，使仪表输出更大的信号。该装置可用于测量导电或非导电的液体，但被测液体的介电常数必须保持恒定。

如果容器内的液体是导电的（图 5-4），那么液体就不能用作电容器的绝缘介质，因此，为导电液体设计的电容式液位探头上涂上塑料或其他一些电介质（作为极板间的电介质）。此时，金属探头构成电容器的一个极板，导电液体构成另一个极板。在这种电容器的设计中，变量是介电常数（ε）和距离（d），由于液面上升，液体取代了低介电常数的气体，并使得罐壁更接近探头（两个极板靠近，d 变小）。这意味着当液位最大时，总电容将达到最大值，此时两极板之间的 ε 达到最大，有效距离 d 最小；当液位最低时，两极板之间就是空气，ε 最小，而极板距离 d 最大，电容达到最小值。

如果液体是不导电的（图 5-5），它可以直接用作电介质，容器的金属壁形成第二电容器板。探头是一根裸露的金属电缆或金属棒。在这种设计中，影响电容量的唯一变量就是介电常数 ε。液体的介电常数大大超过上方气体的介电常数，这意味着当液位最大时，总电容将达到最大值，两极之间充满了介电常数较大的液体；而液位最小时，两极板之间就是空气，电容也达到最小值。假设探头安装在容器的中心，那么两极间的距离（d）对于不导电的液体测量是恒定的，就是容器的半径。

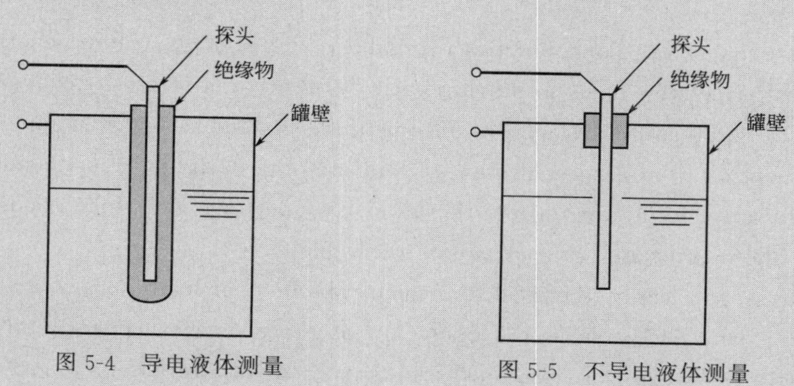

图 5-4　导电液体测量　　图 5-5　不导电液体测量

电容式液位计易于安装。它结构简单，没有活动部件，可用于加压或非加压容器。但是校准很耗时；该单元受材料介电常数变化的影响，因此需要温度补偿；安装时必须确保探头不与罐壁接触；如果应用中需要绝缘探头，在安装过程中必须小心，以免损坏探头的绝缘材料。

5.4 Ultrasonic Level Instruments

Sonic and ultrasonic sensors consist of a transmitter that converts electrical energy into acoustical energy and a receiver that converts acoustical energy back into electrical energy. In sonic sensors, the unit uses the echo principle and emits pulses that have an approximate frequency of 10kHz. After each pulse, the sensor detects the reflected echo. The transmitted and return time of the sonic pulse is relayed electronically and converted into a level indication. The principle for ultrasonic sensors is the same, except that the operating frequency is about 20kHz or higher.

Ultrasonic level instruments (see Figure 5-6) measure the distance from the transmitter (located at some high point) to the surface of a process material located farther below using reflected sound waves. The frequency of these waves extend beyond the range of human hearing, which is why they are called ultrasonic. The time of flight for a sound pulse indicates this distance, and is interpreted by the transmitter electronics as process level. These transmitters may output a signal corresponding either to the fullness of the vessel (fillage) or the amount of empty space remaining at the top of a vessel (ullage).

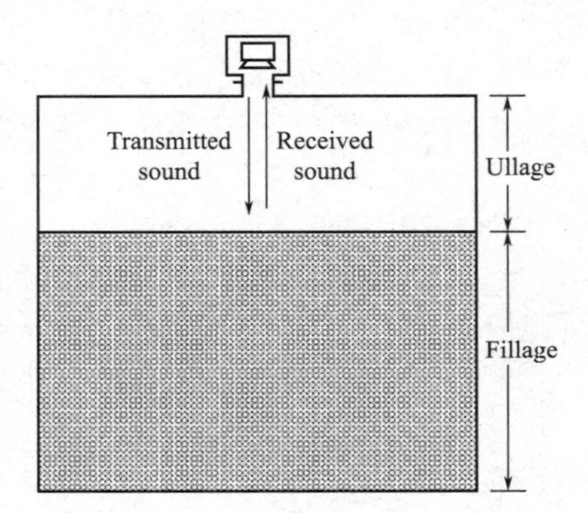

Figure 5-6 Ultrasonic level measurement

Ultrasonic sensors are noncontacting, reliable, and accurate. They penetrate high humidity, are cost effective, have no moving parts, and are unaffected by changes in density, conductivity, or composition. However, strong industrial noise or vibration at the unit's operating frequency will affect performance, and in some designs, dusts tend to give false signals. In addition, coating may affect these devices' performance since deposit buildup on the probe (or the membrane) will attenuate the signal.

Ultrasonic sensors cannot be used to measure the level of foam because the sound signal is absorbed by foam. Also, since the operation of these devices depends on the speed of sound, they will not work in a vacuum. Various factors can affect the speed of sound and so the instrument's accuracy, vapor concentration, pressure, temperature, relative humidity, and the presence of other gases/vapors. Frequently, temperature compensation may be required to avoid variations in accuracy.

【译文】

5.4 超声波液位计

声波和超声波传感器由一个将电能转换成声能的发射器和一个将声能转换回电能的接收器组成。在声波传感器中，该装置使用回声原理并发出频率约为10kHz的脉冲。在每个脉冲之后，传感器检测反射的回波。声波脉冲的发射和返回时间通过电子方式传送并转换为被测量液位的指示值。超声波传感器的原理也是这样，但它的工作频率更高，约为20kHz。

超声波液位计（图5-6）利用反射声波来测量从发射器（位于某一高点）到被测介质表面的距离。这些声波的频率超出了人类的听觉范围，被称为超声波。声音脉冲的飞行时间可表示距离，并由变送装置转换为液位。这些变送器可以输出与容器的填充量（液位值）或容器液位上部的空余量（气相值）相对应的信号。

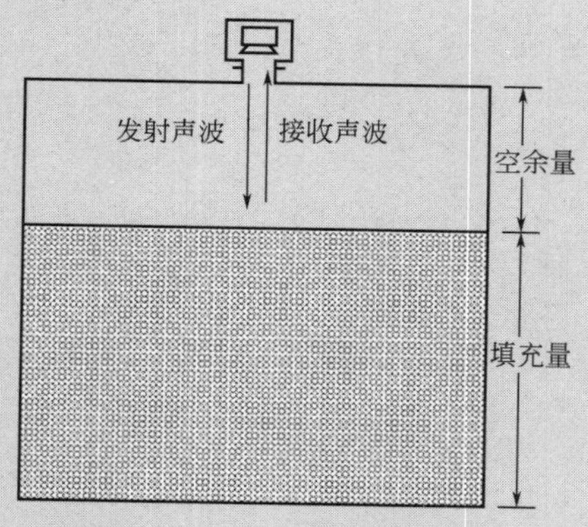

图5-6 超声波液测量

超声波传感器采用非接触测量，可靠准确，能穿透高湿度介质，经济有效，没有活动部件，不受密度、导电性或成分变化的影响。然而，强烈的工业噪声或机组工作频率附近的振动将影响其性能，灰尘在某些应用场合也会引起错误的信号。此外，涂层可能会影响这些器件的性能，因为探头上的沉积物会使信号衰减。

超声波传感器不能用来测量泡沫的液位，因为声音信号会被泡沫吸收。此外，由于这些设备的运行依赖声波，所以无法在真空中工作。有许多因素可以影响声速，从而影响仪表的准确性，例如蒸汽浓度、压力、温度、相对湿度和其他气体/蒸汽的存在。通常，为保证精度，可能需要温度补偿。

Words and Expressions 词汇和短语

1. pharmaceutical /ˌfɑːrməˈsuːtɪkl/ adj. 制药的
2. petroleum /pəˈtroliəm/ n. 石油
3. ultrasonic /ˈʌltrəˈsɔnik/ adj. 超声波的
4. pulse echo 脉冲回波
5. pulse radar 脉冲雷达
6. strain gauge 应变仪

7. gas phase　气相

8. knockout pot　凝液罐

9. zero suppression　零点抑制

10. zero elevation　零点抬升

11. dielectric　/ˌdaiəˈlektrik/　n. 电介质

12. probe　/proʊb/　n. 探头

13. acoustical　/əˈkustɪkl/　adj. 声学的

14. penetrate　/ˈpenətreɪt/　vi. 渗透

15. humidity　/hjuːˈmɪdəti/　n. 湿度

16. deposit　/diˈpɔzit/　vi. 沉淀

Chapter 6

Control Modes

第6章

控制模式

6.1 Introduction

A controller compares its measurement to its setpoint and, based on the difference between them (e=error), generates a correction signal to the final control element (e. g., control valve) in order to eliminate the error. The way a controller responds to an error determines its character, which significantly influences the performance of the closed-loop control system.

Industrial controllers are usually made to produce one, or a combination of, control actions (modes of control). These include

- on-off or two-position control
- proportional control
- proportional plus integral control
- proportional plus derivative control
- proportional plus integral plus derivative (PID) control.

Proportional control (P) is the main and principal method of control. It calculates a control action proportional to the error. Proportional control cannot eliminate the error completely. Integral control (I) is the means to eliminate the remaining error or offset value, left from the proportional action, completely. This may result in reduced stability in the control action. Derivative control (D) is sometimes added to introduce dynamic stability to the control loop.

A controller is called direct acting if its output increases when its measurement rises and is called reverse acting if its output decreases when its measurement rises.

【译文】

6.1 简介

控制器将其测量值与设定值进行比较，根据两者之间的偏差，向控制元件（如控制阀）发出校正信号，以消除偏差。控制器对偏差的响应方式决定了控制器的性质，对闭环控制系统的性能有着重要影响。

工业控制器通常采用一种或多种控制模式的组合。这些控制模式包括：

- 开关或双位置控制；
- 比例控制；
- 比例＋积分控制；
- 比例＋微分控制；
- 比例＋积分＋微分（PID）控制。

比例控制（P）是控制的主要方法，它产生与误差成比例的控制动作，比例控制不能完全消除偏差（存在余差）。积分控制（I）用以消除纯比例控制不能消除的余差，这可能导致控制动作的稳定性降低。微分控制（D）作用的加入，主要是增强控制回路的动态稳定性。

如果一个控制器的输出在其测量值上升时增加，则称为正作用控制器；如果其输出在其测量值上升时减少，则称为反作用控制器。

6.2 On-Off Control

On-off control is a discontinuous form of control action, and is also referred to as two-position control. The technique is crude, but can be a cheap and effective method of control if a fairly large fluctuation of the process variable is acceptable.

A perfect on/off controller is "on" when the measurement is below the setpoint. Under such conditions the manipulated variable is at its maximum value. When the measured variable is above the setpoint, the controller is "off" and the manipulated variable is at its minimum value.

$$\text{if} \begin{cases} e>0 \quad \text{then} \quad m=\max \\ e<0 \quad \text{then} \quad m=\min \end{cases} \tag{6-1}$$

On-off control should only be used where cyclic control is permissible (e. g. , in large-capacity systems). On-off control cannot provide steady measured values, but it is good enough for many applications. Most people are familiar with the technique as it is commonly used in home heating systems and domestic water heaters. Consider the control action on a gas-fired boiler for example. When the temperature is below the setpoint, the fuel is on; when the temperature rises above the setpoint, the fuel is off.

In most practical applications, due to mechanical friction or to the arcing of electrical contacts, the error must exceed a narrow band (around zero error) before a change will occur. This band is known as the differential gap, and its presence is usually desirable to minimize the cycling of the process.

【译文】

6.2 开关控制

开关控制是一种不连续的控制动作，也称为双位控制。这种方法虽然粗糙，但如果过程变量的较大波动可以接受，那么它就是一种廉价而有效的控制方法。

当测量值低于设定值时，理想的双位控制器打开，操纵变量处于最大值；当测量值高于设定值时，双位控制器关闭，操纵变量处于其最小值。

$$\begin{cases} e>0, \ m=\max \\ e<0, \ m=\min \end{cases} \tag{6-1}$$

双位控制只适合在允许循环控制的地方使用（例如大容量系统中）。双位控制不能确保变量的稳定，但它足以满足许多应用，例如人们熟悉的家庭供暖系统和热水器。以某燃气锅炉的控制为例，当温度低于设定值时，燃料阀打开；当温度超过设定值时，燃料阀关闭。

在大多数实际应用中，由于机械或电触点的间隙，误差往往要超过某个范围才会令控制器动作，这个区间被称为间隙带，它的存在通常是为了减少开关控制过程的循环。

6.3 Proportional Control

The proportional action provides an output that is proportional (in linear relation) to the direction and magnitude of the error signal, as shown in Figure 6-1. The adjustable parameter of the proportional mode, K_c, is called the proportional gain. Some controller vendors use the term gain while others use proportional band (PB) to describe a similar function. The relationship between a controller's gain and its PB value is as follows:

$$K_c = \frac{100}{PB} \tag{6-2}$$

"Wide bands" (high PB values) correspond to less "sensitive" controller settings, and "narrow bands" (low PB percentages) correspond to more "sensitive" controller settings.

As the name "proportional" suggests, the correction generated by the proportional control mode is proportional to the error. The larger the gain, the larger the change in the controller output caused by a given error. The proportional controller responds only to the present. It cannot consider the past history of the error or the possible future consequences of an error trend. It simply responds to the present value of the error. Equation (6-3) describes the operation of the proportional controller.

$$m = K_c e + b = \left(\frac{100}{PB}\right) e + b \tag{6-3}$$

where

m = the output signal to the manipulated variable (control valve)

K_c = the gain of the controller

e = the deviation from set point or error

PB = the proportional band

b = the live zero or bias of the output, which in pneumatic systems is usually 20kPa and in analog electronic loops is 4mA.

For example, If the PB is set at 20% ($K_c = 5$) then a change in the PV, or input, from 40% to 60% will result in the same change of the MV, or output, from 0 to 100%. With the same resultant motion of the valve from fully closed to fully open. Likewise, a PB value of 500% ($K_c = 0.2$) will result in the MV, or output, changing from 40% to 60% when the PV, or input, changes from 0 to 100%.

The main limitation of plain proportional control is that it cannot keep the controlled variable on setpoint. It is evident that by increasing the gain, one can reduce (but not eliminate) the offset. If the controller gain is very high, the presence of the offset is no longer noticeable, and it seems as if the controller is keeping the process on setpoint. Unfortunately, most processes become unstable if their controller is set with such high gain. The only exceptions are the very slow processes. For this reason the use of plain proportional control is limited to slow processes that can tolerate high controller gains (narrow proportional bands).

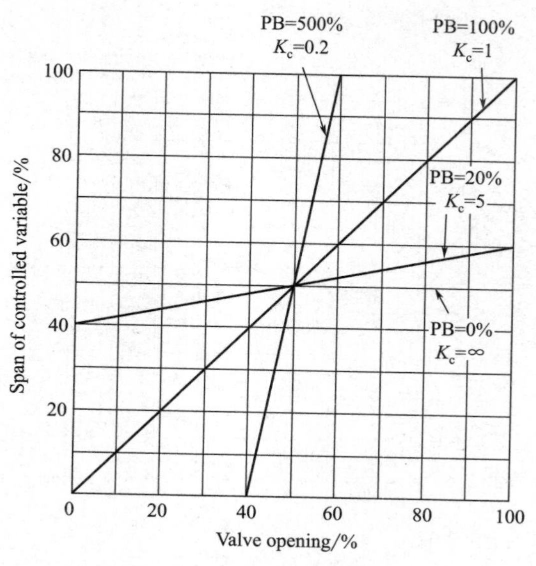

Figure 6-1 K_c and PB

As shown in Figure 6-2, After a permanent load change, the proportional controller is incapable of returning the process back to the setpoint and an offset results. The smaller the controller's gain, the larger will be the offset.

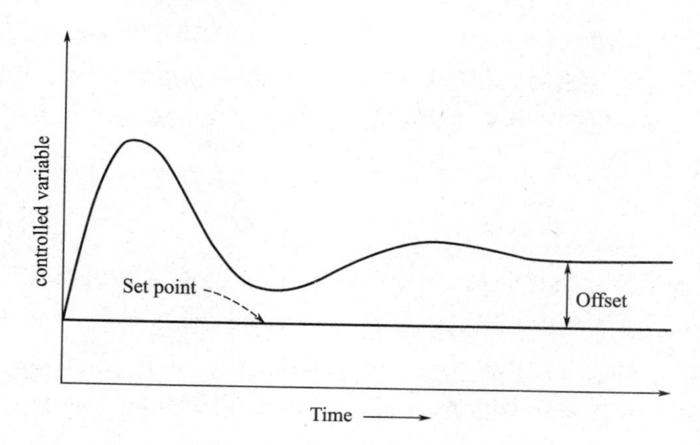

Figure 6-2 Plain proportional control causes offset

【译文】

6.3 比例控制

比例作用可以产生与偏差信号成比例（线性关系）的输出（图 6-1）。比例控制模式的可调参数K_c称为比例增益，另外一些用户习惯于采用比例度 PB 来描述。控制器的比例增益K_c与比例度 PB 之间的关系如下：

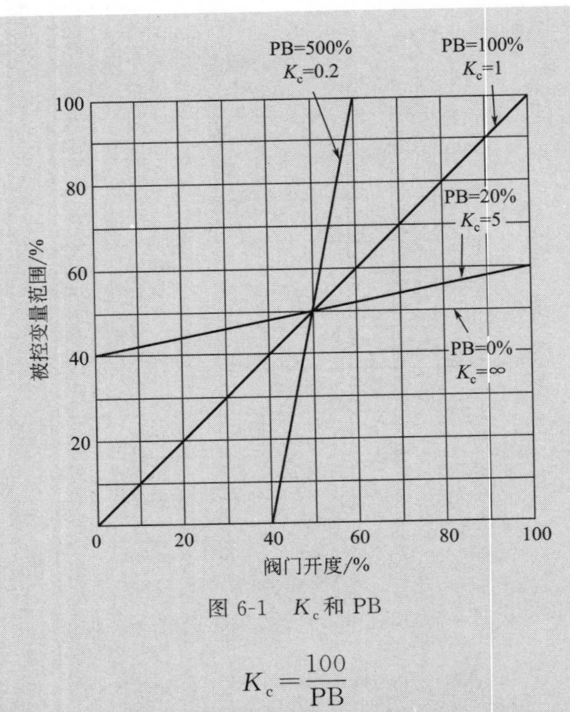

图 6-1 K_c和 PB

$$K_c = \frac{100}{PB} \tag{6-2}$$

宽比例度（PB 值高）表示比例控制作用不强，窄比例度（PB 值低）表示比例控制作用很强。

顾名思义，比例控制模式产生的校正与偏差成正比。比例增益越大，由给定偏差引起的控制器输出的变化就越大。比例控制器只对当前值有效，它不能对偏差的过去或偏差变化的趋势产生控制作用，它只能作用于当前偏差值。式(6-3）描述了比例控制器的输入输出特性：

$$m = K_c e + b = \left(\frac{100}{PB}\right) e + b \tag{6-3}$$

式中，m 是去操纵变量（控制阀）的输出信号；K_c是控制器比例增益；e 是测量值与设定值之间的偏差；PB 是比例度；b 是输出信号的偏置（或初始值），在气动系统中通常为 20kPa 的气压，在模拟量信号回路中为 4mA 的电流。

例如，如果将 PB 设置为 20%（$K_c=5$），则从 40%变为 60%的输入变化将导致从 0%到 100%的输出变化，阀门将从完全关闭到完全打开。同样，如果 PB 为 500%（$K_c=0.2$）时，当输入从 0%变化到 100%时，输出（阀门开度）将从 40%变为 60%。

纯比例控制的主要局限性是不能将被控变量保持在设定值上。很明显，通过增大比例增益，可以减小（但不能消除）偏差。如果控制器比例增益非常高，那么余差就小到可忽略，此时控制器似乎可以使变量保持在设定值上。但事实是，如果控制器的比例增益设置过高，大多数的控制过程都会变得不稳定，除了非常慢的过程。由于这个原因，纯比例控制仅适用于缓慢变化的过程，其可以设置很高的控制器增益（窄比例度）。

在恒定的负载变化后，比例控制器不能使过程返回到设定值并产生余差。控制器比例增益越小，余差越大。如图 6-2 所示。

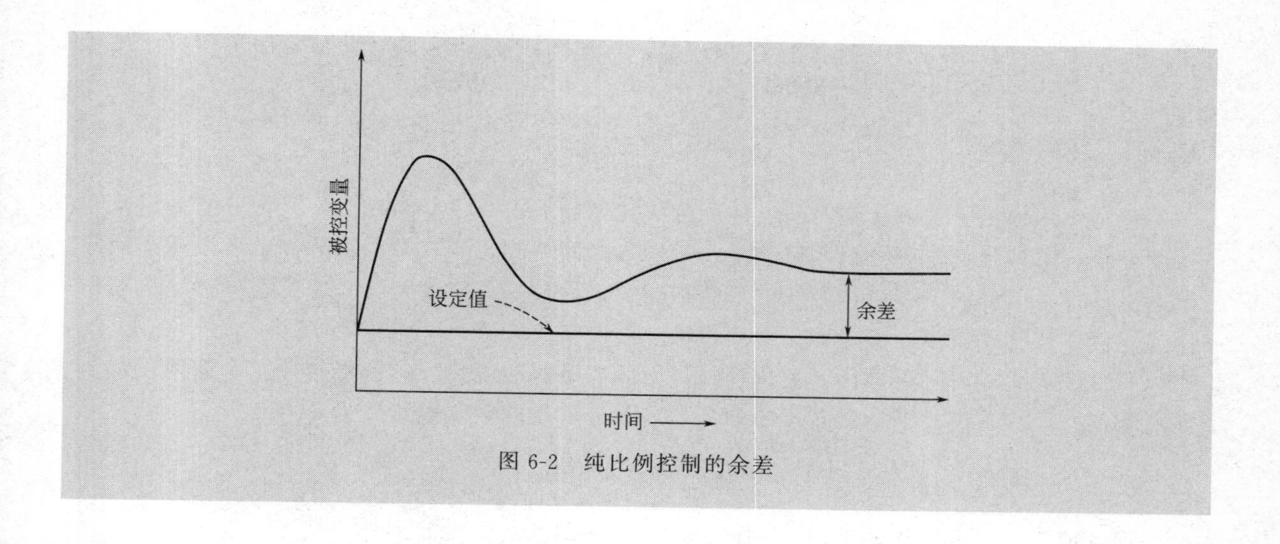

图 6-2 纯比例控制的余差

6.4 Integral Control

The Integral action provides an output that is proportional to the time integral of the input. That is, the output continues to change as long as an error exists. In other words, the integral function acts only when the error exists for a period of time. The integral function is used to gradually eliminate the offset. Loops with low gain only (i. e. , no integral function) will provide stable performance but will generate large offsets, and vice versa. The integral function is slower than the proportional function because it must act over a period of time.

The integral control mode is also sometimes called reset mode because after a load change it returns the controlled variable to setpoint and eliminates the offset, which the plain proportional controller cannot do. The mathematical expression of the integral function is

$$m=\frac{1}{T_{\mathrm{i}}}\int e\,\mathrm{d}t+b \tag{6-4}$$

The integral function is not used alone, the mathematical expression for a proportional plus integral (PI) controller is

$$m=K_{\mathrm{c}}\left[e+\frac{1}{T_{\mathrm{i}}}\int e\,\mathrm{d}t\right]+b \tag{6-5}$$

The term T_{i} is the integral time setting of the controller, also called reset time or repetition time. Its unit is minute, which represents the time of resetting once. With a step input, the output of a pure integrator reaches the value of the step input during the integral time, assuming $b=0$.

One disadvantage of integral control is that it can easily cause the problem of integral saturation. This occur when the deviation cannot be eliminated, for example on an open loop, so the controller is driven to its extreme output. This condition creates loss of control for a period of time, followed by extreme cycling. Implementing protection from such an occurrence is generally necessary and can be built into the controller as an "anti-integral windup."

【译文】

6.4 积分控制

积分作用提供了与输入在时间上的积分成比例的输出。只要存在偏差，积分的输出就会持续变化。也就是说，积分只在偏差存在的一段时间内起作用。利用积分控制可以逐步消除偏差。低增益（如没有积分作用）的控制回路稳定性较好，但偏差较大，反之亦然。积分作用与比例作用相比较慢，因为积分必须作用一段时间。

积分控制模式有时也被称为复位模式，因为在负载改变后，它使得被控变量返回设定值并消除偏差，这是纯比例控制器不能做到的。积分作用的数学表达式为

$$m=\frac{1}{T_i}\int e\,dt+b \tag{6-4}$$

积分控制不单独使用，比例加积分（PI）控制器的数学表达式为

$$m=K_c\left(e+\frac{1}{T_i}\int e\,dt\right)+b \tag{6-5}$$

式中，T_i是控制器的积分时间，它也被称为复位时间或重复时间。积分时间的单位是min，代表复位一次所需的时间。假设当$b=0$，输入是一个单位阶跃信号时，积分控制器的输出将在一个积分时间（T_i）内达到阶跃输入的值。

积分控制的缺点是容易引起积分饱和问题。当偏差无法消除时，就会发生积分饱和状况，例如在开环回路中，积分作用将使控制器的输出达到其极限值。这种情况会造成一段时间内的过程失控，然后是极端的循环。通常必须防止这种情况的发生，可以在控制器中内加入“抗积分饱和措施”。

6.5 Derivative Control

The derivative action provides an output that is proportional to the rate of change (derivative) of error. In other words, the derivative function acts only when the error is changing with time. The derivative speeds up the controller action, compensating for some of the delays in the feedback loop. It is used to provide quick stability to sudden upsets. The mathematical expression of the derivative function is

$$m=T_d\frac{de}{dt}+b \tag{6-6}$$

The derivative function is not used alone, the only purpose of derivative control is to add stability to a closed loop control system. As derivative control on its own has no purpose, it is always used in combination with P control or PI control. This results in a PD control or PID control. The mathematical expression for a proportional plus derivative (PD) controller is

$$m=K_c\left(e+T_d\frac{de}{dt}\right)+b \tag{6-7}$$

The term T_d is the derivative time setting of the controller, its unit is minute. This is the length of time by which the D mode (derivative mode) "looks into the future." The greater the derivative time, the stronger the derivative effect. In other words, if the derivative mode is set for a time T_d, it will generate a corrective action immediately when the error starts changing, and the size of that correction will equal the size of the correction that the proportional mode would have generated T_d time later. The longer the T_d setting, the further into the future the D mode predicts and the larger its corrective contribution. When the slope of the error is positive (measurement is moving up relative to the setpoint), the derivative contribution will also rise, if the controller is direct-acting.

The derivative control became necessary as the size of processing equipment increased and, correspondingly, the mass and the thermal inertia of such equipment also became greater. For such large processes it is not good enough to respond to an error when it has already evolved because the flywheel effect (the inertia or momentum) of these large processes makes it very difficult to stop or reverse a trend once it has evolved. The purpose of the derivative control is to predict the process errors before they have evolved and take corrective action in advance.

【译文】

6.5 微分控制

微分作用提供了与偏差变化率成比例的输出。换句话说，微分作用只在偏差随时间变化的时候才起作用。微分作用加速了控制器的动作，补偿了反馈回路中的一些延迟，它用于突然扰动的快速稳定。微分作用的数学表达式为

$$m=T_{\mathrm{d}}\frac{\mathrm{d}e}{\mathrm{d}t}+b \tag{6-6}$$

微分控制不能单独使用，微分控制的唯一目的是增加闭环控制系统的稳定性。由于纯微分控制本身没有意义，所以通常与比例控制或比例积分控制结合使用，构成一个比例微分控制或比例积分微分控制。比例加微分（PD）控制器的数学表达式为

$$m=K_{\mathrm{c}}\left(e+T_{\mathrm{d}}\frac{\mathrm{d}e}{\mathrm{d}t}\right)+b \tag{6-7}$$

微分控制的主要参数是微分时间 T_d，单位是 min。这是微分作用"展望未来"的时间长度。微分时间越大，微分作用越强。即如果设置一个微分时间 T_d，当偏差开始改变时，微分将立即产生纠错控制，经过 T_d后，微分部分的输出等于比例部分的输出。设置的 T_d越长，微分模式对后期的影响越大，其校正作用也越大。当偏差的斜率为正值时（相对于设定值，测量值在上升），如果控制器是正作用的，那么，微分输出也会上升。

随着过程设备增大，质量和热惯性变得更大时，微分控制就变得很有必要。对于这些惯性较大的控制过程来说，当偏差产生后，很难对其做出很好的响应，因为这些过程的飞轮效应（惯性或动量）导致过程的趋势一旦发生就很难停止或逆转。微分控制的目的是在过程偏差发生变化之前，对其进行预测，并提前采取纠正措施。

6.6 PID Control

Most controllers are designed to operate as PID controllers. When combining the effects of P, I, and D, the typical PID equation is as follows:

$$m = K_c\left(e + \frac{1}{T_i}\int e\,dt + T_d\frac{de}{dt}\right) + b \tag{6-8}$$

If no derivative action is wanted, T_d (derivative time constant) has to be set to zero. If no integral action is wanted, T_i (integral time constant) has to be set to a large value (999min, for example). A very helpful method for understanding the operation of proportional, integral, and derivative control terms is to analyze their respective responses to the same input conditions over time. Figure 6-3 illustrates the responses to a step change in the set point of a pure P, a pure I, and a pure D control mode, and also gives the step responses of the PI, PD, and PID controllers. In each graph, the controller is assumed to be direct-acting (i. e. an increase in process variable results in an increase in output).

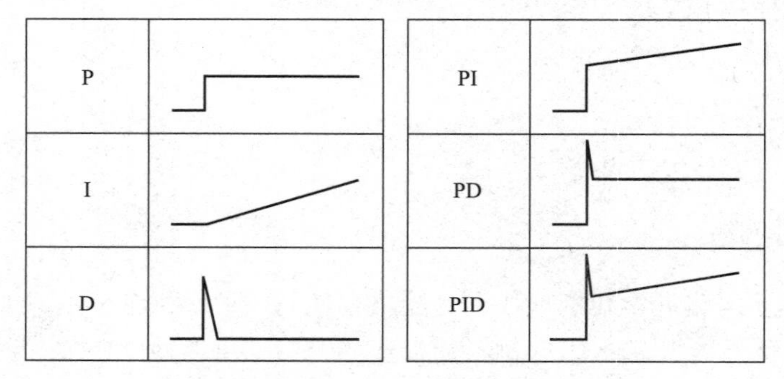

Figure 6-3 Responses to a step change

Figure 6-4 shows the process responses of a closed loop with different control algorithms when they respond to a unit step change in set point. It can be seen that the integral mode returns the process to setpoint while the derivative amplifies transients.

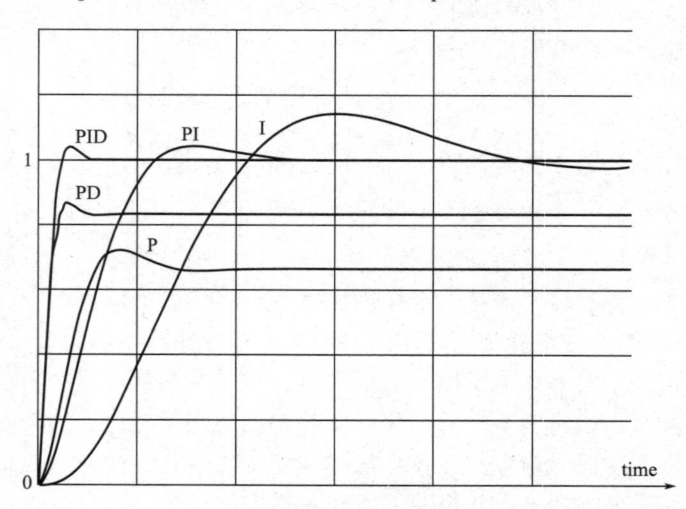

Figure 6-4 The unit step responses of P, I, PI, PD, and PID controllers

The following general rules provide an idea of the PID requirements for different loops. However, keep in mind that each application has its own needs.

- Flow control: P and I are required; D is set at 0 or at minimum.
- Level control: P is required, I is sometimes required, and D is set at 0 or at minimum.
- Pressure control: P and I are required; D is generally set at 0 or at minimum.
- Temperature control: P, I, and D are required, and the integral action is sometimes fairly long.

【译文】

6.6 PID 控制

大多数控制器被设计成 PID 控制器。结合比例、积分、微分作用的影响，典型的 PID 控制方程为：

$$m = K_c\left(e + \frac{1}{T_i}\int e\,dt + T_d\frac{de}{dt}\right) + b \tag{6-8}$$

如果不需要微分作用，微分时间 T_d 设为零。如果不需要积分作用，积分时间 T_i 设置为较大的值（例如 999min）。理解比例控制、积分控制、微分控制特性的一个有效方法，是分析它们在相同输入条件下响应随时间的变化。图 6-3 给出了设定值发生阶跃变化时，纯比例、纯积分、纯微分控制模式的响应，并给出了比例-积分、比例-微分和比例-积分-微分的阶跃响应。在每个图中，都假定控制器是正作用的（即过程变量的增加导致输出的增加）。

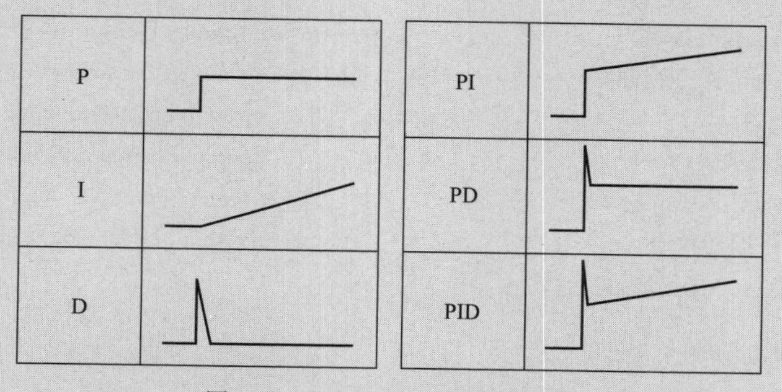

图 6-3 发生阶跃变化时的响应

图 6-4 给出了设定值发生阶跃变化时，采用不同控制算法控制系统的闭环响应过程。可以看出，积分作用将过程拉回到设定值，而微分控制加快了瞬态响应。

下面是不同控制回路对 PID 控制的不同设置需求，是通用规则但并不绝对。

- 流量控制：要求比例和积分作用，微分设为 0 或最小值。
- 液位控制：比例是必需的，积分按需加入，微分作用设置为 0 或最小值。
- 压力控制：要求比例和积分作用，微分作用设置为 0 或最小值。
- 温度控制：要求比例、积分、微分作用，积分动作有时非常长。

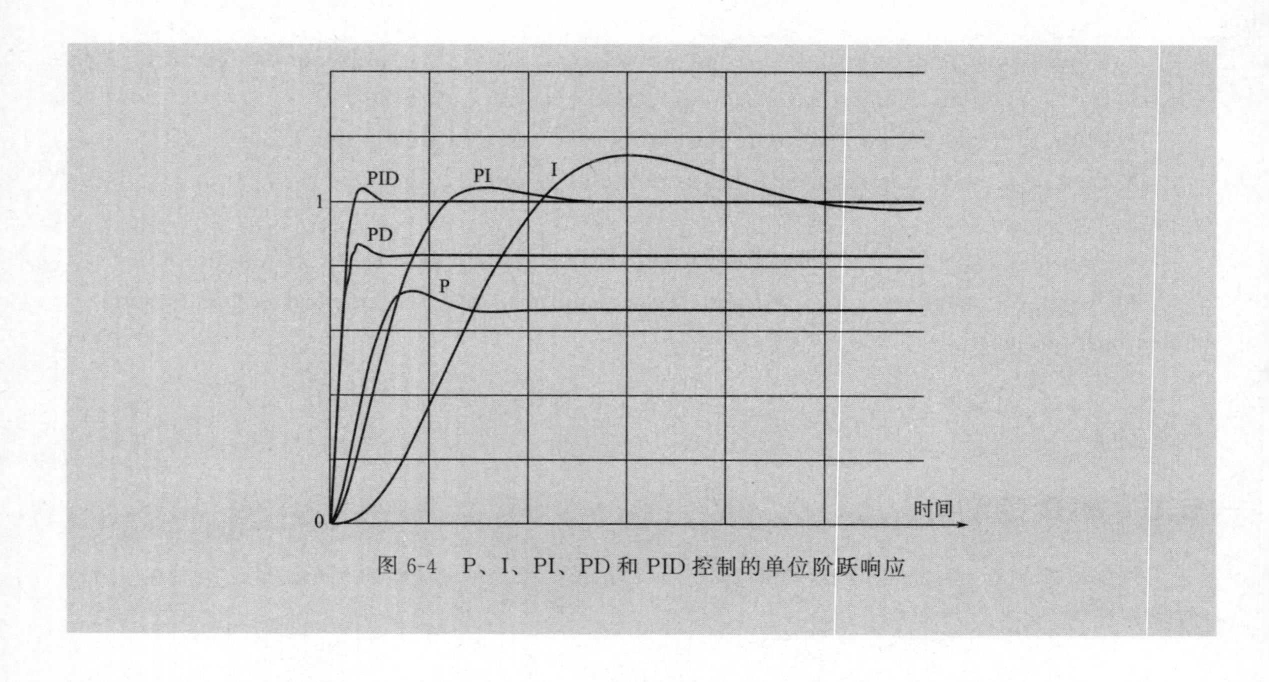

图 6-4 P、I、PI、PD 和 PID 控制的单位阶跃响应

Words and Expressions 词汇和短语

1. pharmaceutical /ˈfɑːrməˈsuːtɪkl/ adj. 制药的
2. controller /kənˈtroulər/ n. 控制器
3. setpoint /ˈsetpɔint/ 设定值
4. proportional control 比例控制
5. integral control 积分控制
6. derivative control 微分控制
7. direct act 正作用
8. reverse act 反作用
9. fluctuation /ˌflʌktʃuˈeɪʃn/ n. 起伏，波动
10. cyclic control 循环控制
11. proportional gain 比例增益
12. proportional band 比例度
13. permanent /ˈpɜːrmənənt/ adj. 永久的
14. offset /ˈɔːfset/ n. 余差
15. reset time 积分时间
16. integral saturation 积分饱和
17. extreme /ɪkˈstriːm/ adj. 极端的
18. upset /ʌpˈset/ n. 扰动
19. derivative time 微分时间
20. predict /prɪˈdɪkt/ v. 预测
21. flywheel effect 飞轮效应

Chapter 7

Control Valves

第 7 章

控制阀

7.1 Introduction

A Control valve is a power-operated device used to modify the fluid flow rate in a process system, it works to restrict the flow through a pipe at the command of a remotely sourced signal, such as the signal from a loop controller or logic device (such as a PLC), or even a manual interface controlled by a human operator. Control valves are comprised of two major parts: the valve body, containing all the mechanical components necessary to influence fluid flow; and the valve actuator, providing the mechanical power necessary to move the valve body components.

In most applications, a control valve is the final element in a control loop. Control valve maintain process variables such as pressure, flow, temperature, or level at its desired value, despite changes in process dynamics and load. Of the three basic components of a typical control loop (sensor, controller, and valve), the valve is subject to the harshest conditions and is the least understood. To complicate matters, the valve is also the most expensive and most likely to be selected incorrectly.

When choosing a control valve for a process, there are many things that must be considered. Selecting the right valves involves the following factors:

- Process requirements The type of fluid passing through the valve, the inlet pressure and differential pressure across the valve, the maximum and minimum flows, the flowing temperature, and the degree of shutoff.
- Correct sizing of the valve The valve must be able to handle its maximum design flow. However, the designer must avoid oversizing or undersizing since they degrade the valve's operation. Typically, a properly sized valve should not operate below the 10 percent or above the 90 percent travel position.
- Suitable flow characteristics The valve's flow characteristics must match the process requirements (i. e. , linear, equal percentage, or quick-opening).
- Fail-safe mode (on air and/or signal failure) An air-to-open valve is a fail-close valve (FC), an air-to-close valve is a fail-open valve (FO).
- For toxic or environmentally harmful applications, the valve body type and accessories must be selected correctly.
- Always refer to the vendor's recommendations for proper installation.

【译文】

7.1 简介

控制阀是一种动力操作装置，用于改变控制过程中的流体流量，它根据外部信号来限制通过管道的流量，例如来自回路控制器、逻辑设备（例如PLC）的信号，甚至是人工操作的命令。控制阀由两个主要部分组成：阀体，包含调节流体流量所需的所有机械部件；阀门执行机构，提供必要的机械动力来移动阀体组件。

在生产应用中，控制阀是控制回路中的最终执行元件。不管是过程波动或负载变化导致压力、流量、温度和液位等发生变化，都需要通过控制阀将这些过程变量保持在期望值上。在典型控制回路的三个基本组成部分（传感器、控制器和阀门）中，阀门所处的环境最为恶劣，人们对其了解也最少，阀门也是其中最贵和最容易被选错的。

在为控制过程选择阀门时，必须认真考虑各种因素。选择合适的阀门涉及以下因素。

- 工艺要求　通过阀门的流体类型、进口压力和压差、最大流量和最小流量、流体温度和关闭程度。
- 正确的阀门口径　阀门必须能够承受最大的设计流量，但是应避免口径过大或过小，这样会降低阀门的运行效率。通常，适当大小的口径应该保证阀门不会在10%以下或90%以上开度长期运行。
- 合适的流量特性　阀门的流量特性必须符合工艺要求（即线性、等百分比或快开等）。
- 故障安全模式（空气和/或信号故障时）　气开式阀门是一种故障关闭式（FC）阀门，气关式阀门是一种故障开启式（FO）阀门。
- 对于有毒或对环境有害的应用场合，必须正确选择阀体类型和附件。
- 正确的安装方式必须严格按照阀门厂家的建议。

7.2 Valve Bodies

Valve bodies can be classified into two types based on their motion: linear and rotary. Linear valves, also known as multi-turn valves, have a sliding-stem design that pushes a closure element into an open or closed position. Linear valves comprise globe, diaphragm, or pinch valves.

Rotary valves are also known as quarter-turn valves, quarter-turn valves will be in their fully open or fully closed state after a 90° turn of the stem, they run much faster than linear valves. Rotary valves include ball, butterfly, and plug, etc.

Each type of valve has its special generic features, which may, in a given application, be either an advantage or a disadvantage. Control valves are available with a wide variety of valve bodies in various styles, materials, connections, and sizes. Selection is primarily dependent on the service conditions, the task, and the load characteristics of the application.

(1) Globe valves

Globe valves are the most versatile of all valves, the majority of throttling control valves were of the globe type, characterized by linear plug movements and actuated by spring/diaphragm operators. The main advantages of the globe valve include the simplicity of the spring/diaphragm actuator, low cavitation and noise, a wide range of designs for corrosive, abrasive, high temperature and high pressure applications, a linear relationship between the control signal and valve stem movements and relative small amounts of dead band and hysteresis values.

Globe valves are available with either single-or double-seated construction, and may have pressure-balanced trim.

Single-seated valves (see Figure 7-1) are the most widely used of the globe body patterns. They are available in a wide variety of configurations, including special-purpose trims. They have good seating shut-off capability, are less subject to vibration due to reduced plug mass, and are generally easy to maintain. The single-seated valve is suitable for the occasions that require small leakage and small pipe diameter and low pressure difference.

Double-seated valve (see Figure 7-2) requires fewer actuator forces and is top and bottom guided. In a double-seated valve, there will be two opposed force vectors, one generated at the upper plug and another generated at the lower plug. If the plug areas are approximately equal, then the forces will likewise be approximately equal and therefore nearly cancel. This makes for a control valve that is easier to actuate. However, it is more expensive; more difficult to service, maintain, and adjust; and does not provide tight shutoff.

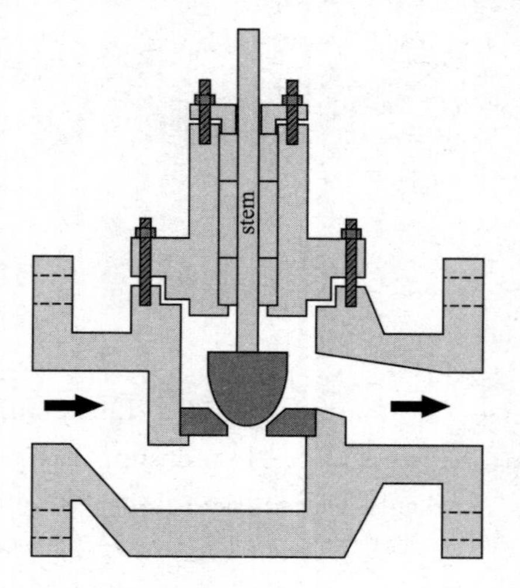

Figure 7-1 Single-seated valve

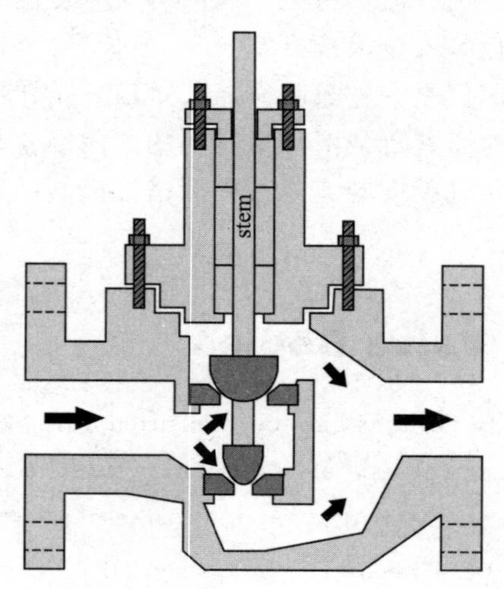

Figure 7-2 Double-seated valve

Three-way valves are a specialized double-seated globe valve configuration. There are two basic types. One is for mixing service, i. e. , the combination of two fluid streams passing to a common outlet port. The other is for diverting service, i. e. , taking a common stream and splitting it into two outlet ports. Angle valves are mainly used to change the direction of fluid flow, their inlet and outlet are perpendicular to each other, instead of going straight in and out like ball valves. Three-way and angle valves are both globe valves.

(2) Diaphragm valves

Diaphragm valves (see Figure 7-3) also known as Saunders valves, they are opened and closed by moving a flexible or elastic diaphragm toward or away from a weir. The elastic diaphragm is moved toward the weir by the pressure applied by a compressor element on the diaphragm. The compressor is connected to the valve stem, which is moved by the actuator. The diaphragm, which at its center is attached to the compressor, is pulled away from the weir when the compressor is lifted. This type valve can be considered as a half pinch valve as only one diaphragm is used, moving relative to a fixed weir. The normal diaphragm valve has a body with side section in the form of an inverted U shape, with the diaphragm closing the orifice at the top.

Diaphragm valves are excellent for sanitary and slurry service as well as for liquids that contain solids or dirt. They are made of a packless construction (because fluid contacts only the liner) and are available as tight shutoff. Diaphragm valves are low-cost devices and their maintenance is simple. In addition, the diaphragm materials available are limited, and due to their application and construction, diaphragm valves tend to be a high-maintenance item.

(3) Ball valves

In the ball valve (see Figure 7-4) design, a spherical ball with a passageway cut through the center rotates to allow fluid more or less access to the passageway. When the passageway is parallel to the direction of fluid motion, the valve is wide open; when the passageway is aligned perpendicular to the direction of fluid motion, the valve is fully shut.

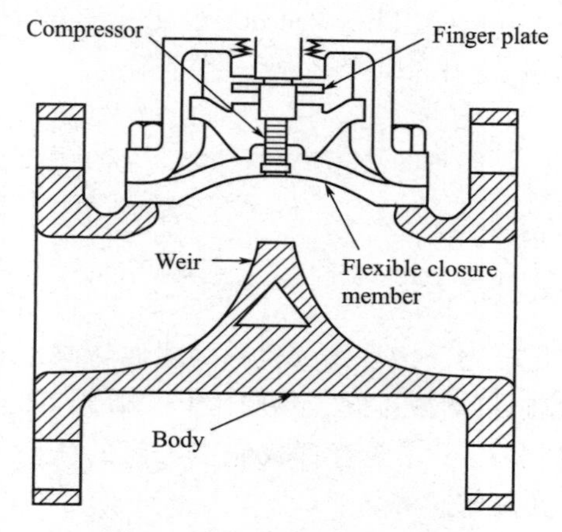

Figure 7-3 Diaphragm valve

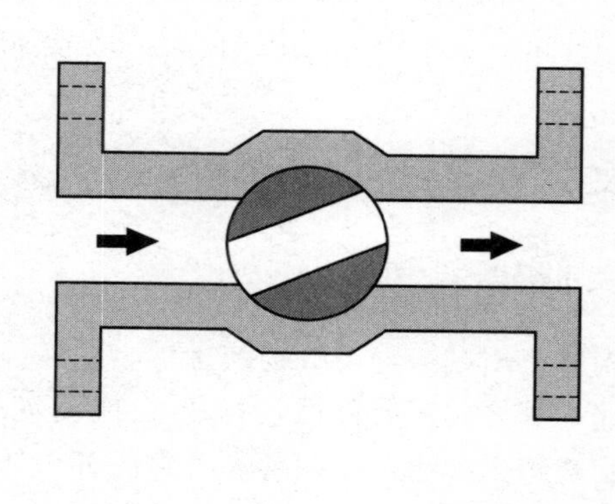

Figure 7-4 Ball valve

Ball valves are good for slurries and fluids with suspended solids because of their straight-through design and self-cleaning action. They have excellent packing sealability, low weight, a small number of parts, and a simple design. With these valves, the valve stem is not alternately wetted and exposed to air (due to the rotary motion), which minimizes the effect of corrosion. In addition, ball valves can provide tight shutoff (with PTFE seals).

(4) Butterfly valves

The butterfly valve (see Figure 7-5) consists of a cylindrical body with a disk (the closure member) mounted on a shaft that rotates perpendicularly to the axis of the valve body. It acts as a damper or as a throttle valve in a pipe and consists of a disk turning on a diametral axis. Like the ball valve its actuation rotation from fully closed to fully open is 90°. Butterfly valves have large capacities, and the only obstruction to the flow is the closure member disk.

The most common butterfly design is the flangeless (wafer) type. Butterfly valves offer high capacity at low cost. They have a small body mass (and so weigh little), are easy to in-

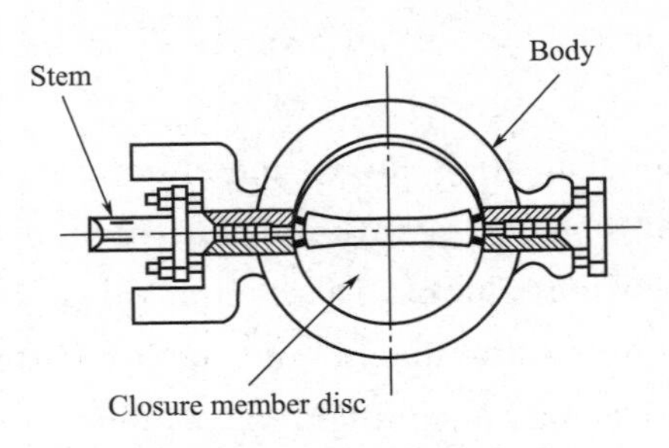

Figure 7-5 Butterfly valve

stall. In addition, they are simply designed and have a small number of parts. The butterfly valve's design eliminates the valve stem's alternate wetting and exposure to air, which minimizes corrosion. These valves have a tight shutoff capability (if they are lined) and can be used as a modulating control valve. However, butterfly valves have a limited pressure-control range, and their pressure-drop ratings are limited. They are not used in cavitation or noise applications or for slurries or dirty/solid-bearing fluids.

【译文】

7.2 阀体

阀体根据其运动可分为两种类型：线性阀和旋转阀。线性阀（也称为多回转阀）具有滑杆设计，可将关闭元件推入打开/关闭位置。线性阀包括截止阀、隔膜阀和夹管阀等。

旋转阀也称为1/4回转阀，在阀杆旋转90°后将处于完全打开或完全关闭的状态，它们的操作比线性阀快得多。旋转阀包括球阀、蝶阀和旋塞阀等。

每种类型的阀门都有其特殊特征，在特定的应用中，这可能是优点也可能是缺点。控制阀可以配置各种类型、材料、连接方式和通径的阀体，选择原则主要取决于阀门的工作条件和应用的负载特性等。

(1) 截止阀

截止阀是所有阀门中用途最广泛的，大多数节流控制阀都是截止阀类型，其特点是由弹簧/膜片机构驱动线性阀塞运动。截止阀的主要优点：弹簧/膜片执行器简单，低气蚀和噪声性，适用于腐蚀、磨蚀、高温和高压等多种应用场合，控制信号和阀杆运动之间呈线性关系，死区和滞后值相对较小。

截止阀可设计成单阀座结构或双阀座结构，并且配置有压力平衡内件。

单阀座阀门（图7-1）是应用最多的截止阀结构型式，有各种配置，包括特殊用途的阀内件。它们具有良好的阀座关闭能力，由于阀芯质量小，较少受到振动的影响，易于维护。单座阀适用于要求泄漏量小、管径小和阀前后压差较低的场合。

双阀座阀门（图7-2）需要较少的执行机构，由顶部和底部导向。在双阀座结构中，有两个相反的压差作用力，一个在上阀芯处产生，另一个在下阀芯处产生。如果阀芯面积大致相等，那么作用力也大致相等，可以相互抵消，这使得控制阀更容易动作。但双阀座更贵，维护和调整的工作量大，不能严密地关闭。

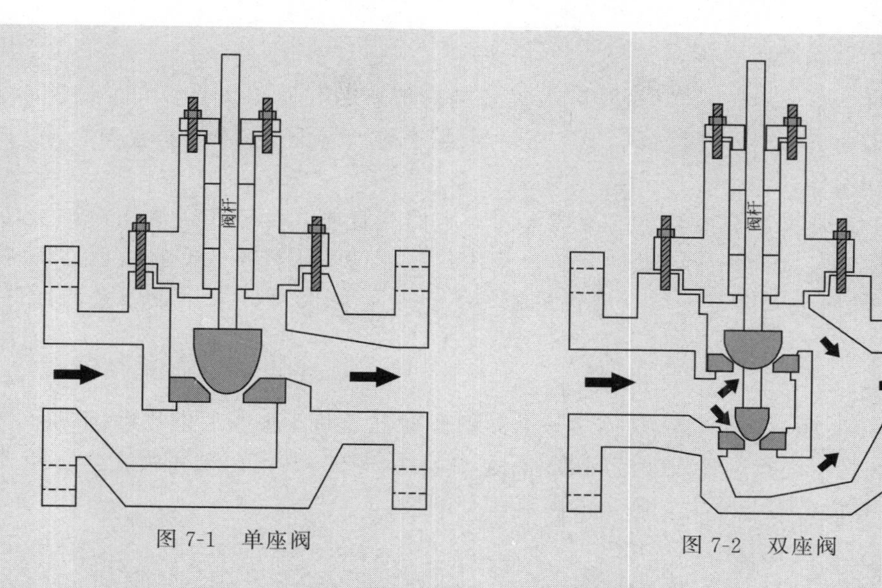

图 7-1 单座阀

图 7-2 双座阀

三通阀是一种专门的双座截止阀结构。有两种基本类型：一个是合流型，即两股流体流过一个共同的出口；另一种是分流型，将一股流体拆分为两个出口流出。角阀主要用于改变流体的流动方向，流体入口和出口的方向彼此垂直，而不像球阀那样直接进出。三通阀和角阀均为截止阀。

（2）隔膜阀

隔膜阀（图 7-3）也称为桑德斯（Saunders）阀，通过将柔性或弹性膜片靠向或移开堰形阀座来打开和关闭。弹性膜片通过压缩机元件在膜片上施加的压力向堰移动。压缩机连接到阀杆，阀杆由执行器驱动。提起压缩机时，将隔膜中央位置处固定在压缩机上的隔板拉开。这种阀可以看做是半夹管阀，因为仅使用一个隔膜相对于固定的堰进行移动。普通隔膜阀的阀体，侧面部分呈倒 U 形，通过隔膜来关闭顶部的流通孔。

隔膜阀适用于清洁流体和泥浆，也适用于含有固体或污物的流体。它们采用无填料结构（因为流体只与内胆接触），可用作严密的关闭阀。隔膜阀是低成本设备，维护简单。此外，隔膜材料有限，由于它们的应用场合和结构，隔膜阀往往后期维护成本高。

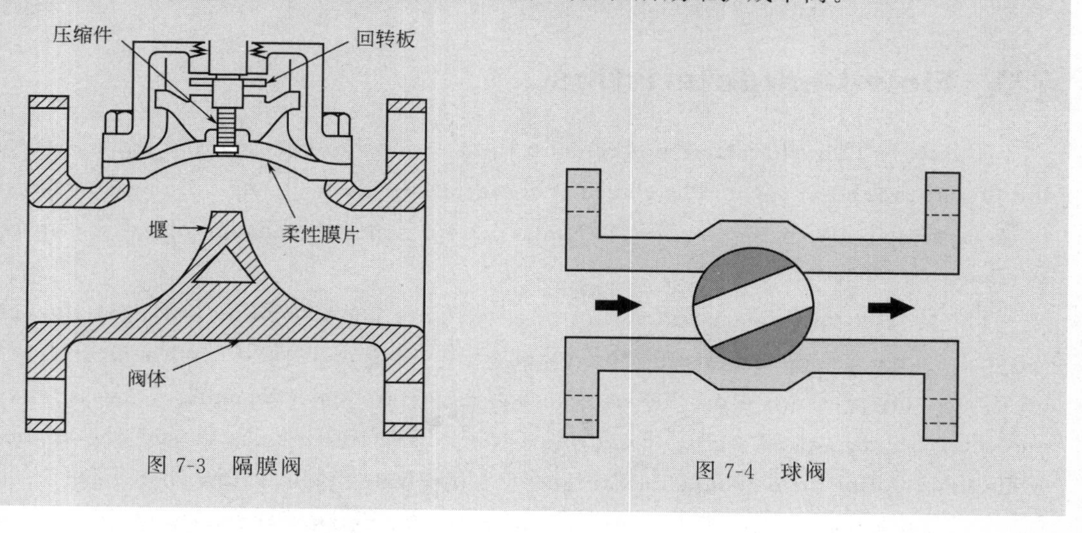

图 7-3 隔膜阀

图 7-4 球阀

(3) 球阀

在球阀（图 7-4）设计中，一个中间有通道的球体旋转，使流体或多或少地进入通道。当通道与流体运动方向平行时，阀门全开；当通道与流体运动方向垂直时，阀门完全关闭。

球阀适用于泥浆和带有悬浮物的流体，因为它们有直通式设计和自清洗功能，具有良好的密闭性，重量轻，部件少，设计简单。使用这些阀门，阀杆不会交替接触空气（因为是旋转运动），从而最大限度地减少腐蚀的影响。此外，球阀可以采用聚四氟乙烯等材料密封。

(4) 蝶阀

蝶阀（图 7-5）由一个圆柱形阀体和一个安装在垂直于阀体轴线上的旋转圆盘（关闭部件）组成。它在管道中起风门或节流阀的作用。像球阀一样，从完全关闭到完全打开只需要驱动旋转 90°。蝶阀的容量很大，唯一的障碍就是阀瓣。

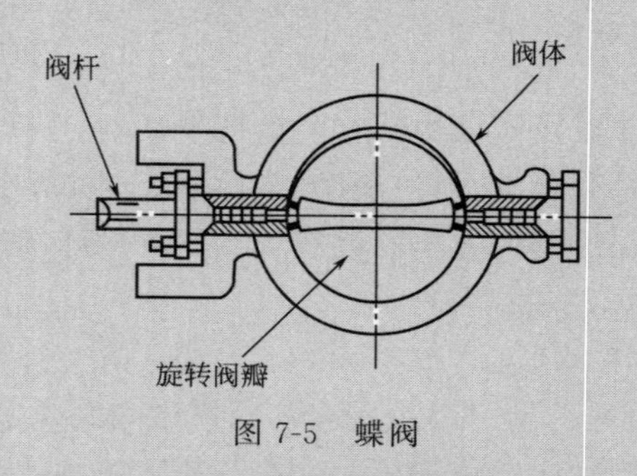

图 7-5　蝶阀

最常见的蝶阀结构是无法兰（对夹式）蝶阀。蝶阀性价比高，重量轻，易于安装。此外，它们设计简单，零件数量少。蝶阀的设计避免了阀杆的交替干湿和暴露在空气中，最大限度地减少了腐蚀。蝶阀具有紧密的关闭能力，可以用作调节阀。但蝶阀的压力控制范围有限，压降额定值也有限。它们不用于空穴或噪声应用，也不能用于泥浆或固体污物的流体。

7.3 Flow Characteristics

A valve's flow characteristics refer to the relationship between the stem position and the flow through the valve. The choice of characteristic has a strong influence on the stability or controllability of the process, because it represents the change of valve gain relative to travel.

The theoretical flow characteristics of a valve are known as the "inherent flow characteristics". They are determined by the design of the plug under test conditions and are based on a constant pressure drop. Typical characteristics are quick opening, linear, and equal percentage. Some valves are available in a variety of characteristics to suit the application, while others offer little or no choice. Figure 7-6 illustrates typical flow characteristic curves.

The quick-opening flow characteristic provides for maximum change in flow rate at low

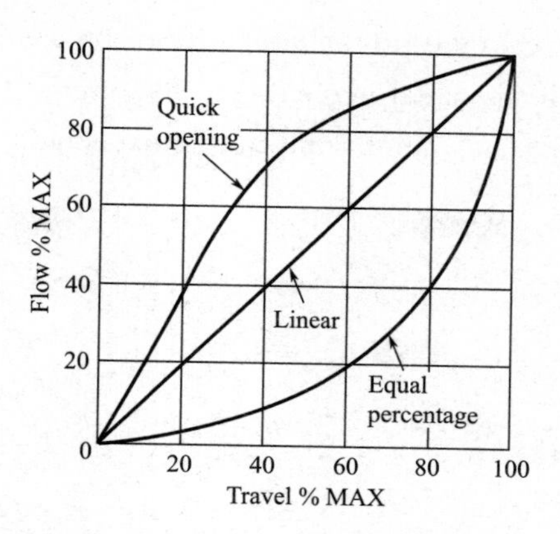

Figure 7-6　Flow characteristic curves

valve travels with a fairly linear relationship. Additional increases in valve travel give sharply reduced changes in flow rate, and when the valve plug nears the wide open position, the change in flow rate approaches zero. In a control valve, the quick-opening valve plug is used primarily for on-off service, but it is also suitable for many applications where a linear valve plug would normally be specified.

The linear valve has a flow rate that varies linearly with the position of the stem. This relationship can be expressed as follows:

$$\frac{Q}{Q_{max}}=\frac{X}{X_{max}} \tag{7-1}$$

where

Q=flow rate

Q_{max}=maximum flow rate

X=stem position

X_{max}=maximum stem position

This proportional relationship produces a characteristic with a constant slope so that with constant pressure drop, the valve gain will be the same at all flows. The linear valve plug is commonly specified for liquid-level control and for certain flow control applications requiring constant gain.

In the equal-percentage flow characteristic, equal increments of valve travel produce equal percentage changes in the existing flow, the change in flow rate is always proportional to the flow rate just before the change in valve plug position is made. When the flow is small, the change in flow rate will be small; With a large flow, the change in flow rate will be large. Generally, this type of valve does not shut off the flow completely in its limit of travel. Thus, Q_{min} represents the minimum flow when the stem is at one limit of its travel. At the fully open position, the control valve allows a maximum flow rate, Q_{max}. So we define a term called rangeability (R) as the ratio of maximum flow (Q_{max}) to minimum flow (Q_{min}):

$$R=\frac{Q_{max}}{Q_{min}} \tag{7-2}$$

Valves with an equal-percentage flow characteristic are generally used on pressure control applications. Valves with an equal-percentage characteristic should also be considered where highly varying pressure drop conditions can be expected.

【译文】

7.3 流量特性

阀门的流量特性是指阀杆位置和通过阀门的流量之间的关系。特性的选择对过程的稳定性或可控性有很大的影响，因为它代表了阀门增益相对于行程的变化。

阀门的理论流量特性称为“固有流量特性”，是由阀芯的设计决定的，并基于恒压降的测试条件。典型的流量特性是快开型、线性、等百分比型。一些阀门具有多种特性以满足应用，而另一些阀门几乎没有选择。图 7-6 展示了典型的流量特性曲线。

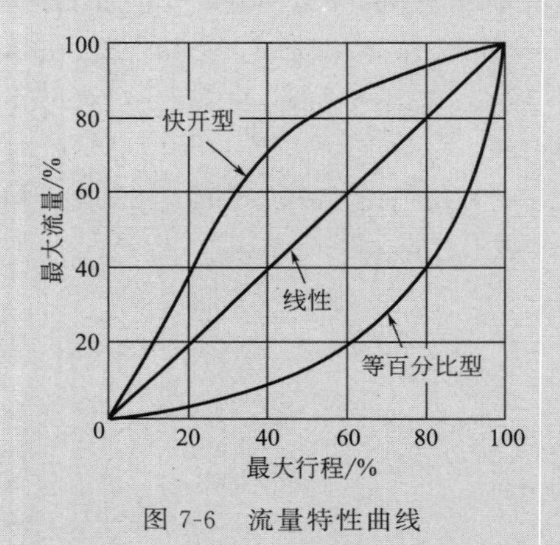

图 7-6　流量特性曲线

快开流量特性的阀门在小开度时，很小的阀杆行程就会令阀门开度（流量）以近似线性方式接近最大值。随着阀杆行程的继续增加，流量的变化急剧减小，当阀芯接近全开位置时，流量的变化接近于零。在控制应用中，快开阀门主要用于开关控制，但它也适用于通常指定为线性阀芯的许多应用场合。

线性阀门具有随阀杆位置线性变化的流量。这种关系可以表示为

$$\frac{Q}{Q_{max}}=\frac{X}{X_{max}} \tag{7-1}$$

式中，Q 是实际流量；Q_{max} 是最大流量；X 是阀杆位置；X_{max} 阀杆最大位置。

这种比例关系产生了一个恒定斜率的特性，在阀门压降恒定的情况下，阀门增益在所有流量下都是相同的。线性阀门通常用于液位控制和需要恒定增益的某些流量控制应用中。

在等百分比流量特性中，相等的阀杆行程增量会在现有流量基础上产生等百分比的变化，流量变化始终与阀芯位置变化之前的流量成正比。当流量较小时，流量的变化率也很小；流量大时，流量的变化率也很大。通常，这种阀不会在其行程极限内完全切断流量，因此，Q_{min} 表示阀杆处于其行程的一个极限时的最小流量。在全开位置，控制阀允许最大流量 Q_{max}。因此，我们将可调比 R 定义为最大流量（Q_{max}）与最小流量（Q_{min}）之比：

$$R=\frac{Q_{max}}{Q_{min}} \tag{7-2}$$

具有等百分比流量特性的阀门通常用于压力控制。在压降条件变化很大的情况下，也可考虑使用等百分比特性的阀门。

7.4 Actuators and Fail-Safe Mode

The actuator provides power to change the orifice area of the valve (i. e., open and close the valve) in response to the received signal. There are two common types of actuators: electric and pneumatic. Electric motor actuators are used to control the opening and closing of smaller rotary-type valves such as butterfly valves. However, pneumatic actuators are used more widely because they can effectively translate a small control signal into a large force or torque.

Proper selection involves process know-ledge, valve knowledge, and actuator knowledge. A control valve can perform its function only as well as the actuator can handle the static and dynamic loads placed on it by the valve. Proper selection and sizing are very important. The following parameters must be known at the beginning of the selection process.

- Power source availability
- Fail-safe modes
- Torque or thrust requirements (actuator capability)
- Control functions
- Economics
- Size, modular construction, easy maintenance.

The most popular and widely used control valve actuator is the pneumatic spring and diaphragm style, see Figure 7-7. Air pressure required to motivate a pneumatic actuator may come directly from the output of a pneumatic process controller, or from a signal transducer (or converter) translating an electrical signal into an air pressure signal. Such transducers are commonly known as I/P converters, since they typically translate an electric current signal of 4 to 20mA DC into an air pressure signal of 20 to 100kPa. Diaphragm actuators are extremely simple and offer low cost and high reliability. Many designs offer either adjustable springs or wide spring selections to allow the actuator to be tailored to the particular application. Because diaphragm actuators have few moving parts that might contribute to failure, they are extremely reliable. Should they ever fail, maintenance is extremely simple.

When choosing a valve, fail-safe mode must be considered. The valve application engineers must choose between fail-close valve (FC) and fail-open valve (FO), the choice will be based upon process safety considerations in the event of control valve air failure.

An air-to-open valve is a fail-close valve (FC), a spring closes the valve on air failure. An air-to-close valve is a fail-open valve (FO), a spring opens the valve on air failure.

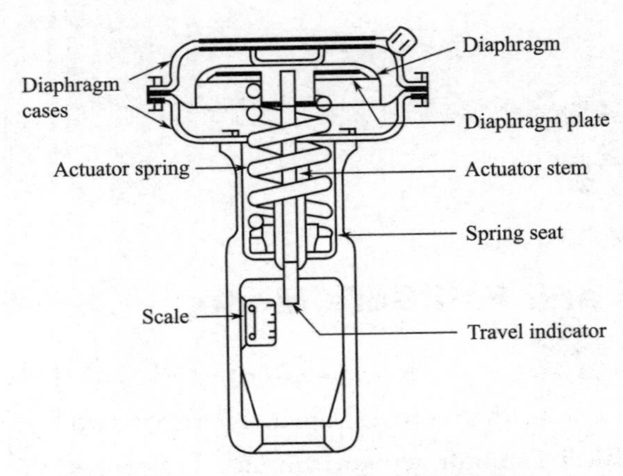

Figure 7-7　Spring and diaphragm actuator

Fail-safe mode involves the selection of actions of actuator and inner valve. Both actuator and inner valve usually offer a choice of increasing air pressure to push the stem down or up, and pushing the stem down may open or close the inner valve. The proper choice of combinations may be made by fail-safe considerations. The process application of the valve must be investigated to determine whether, on instrument air failure, it would be better to have the valve go fully open, fully closed, or remain in its last position. Figure 7-8 summarizes the available valve Fail-safe modes.

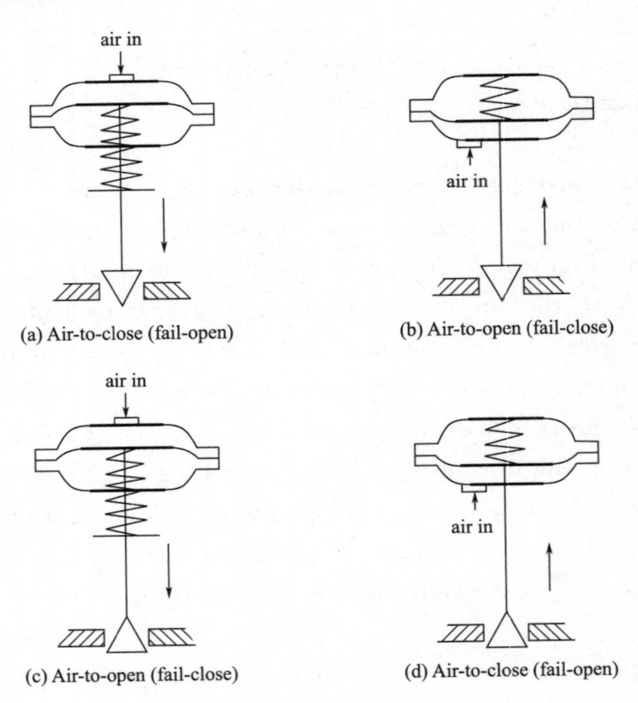

Figure 7-8　Fail-safe mode

For example, consider this automated cooling system for an engine. It is more hazardous to the engine for the valve to fail closed than it would be for the valve to fail open. If the valve is fail-close type, the engine will surely overheat from lack of cooling. If the valve is fail-open type (see Figure 7-9), the engine will merely run cooler than designed, the only negative consequence being decreased effciency. With this in mind, the only sensible choice for a control valve is one that fail-open (air-to-close).

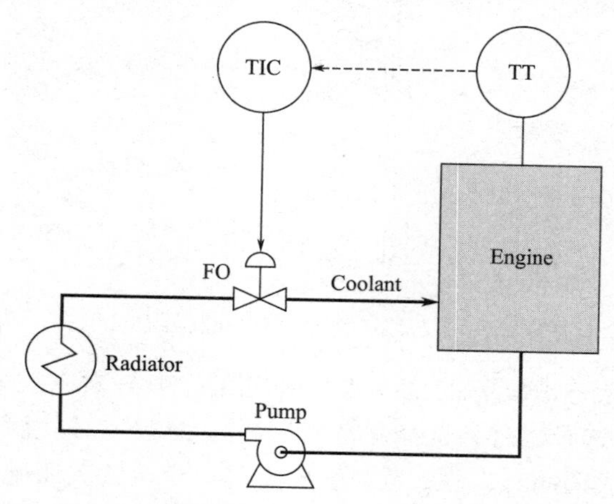

Figure 7-9 Choice of fail-safe mode

【译文】

7.4 执行器与故障模式

执行器根据接收的控制信号产生动力来改变阀门的流通面积（即开大或关小阀门）。执行器有两种常见类型：电动和气动。电动执行器用于控制较小的旋转式阀（例如蝶阀）的打开和关闭；气动执行器可以有效地将较小的控制信号转换为较大的力或转矩，因此使用更为广泛。

执行器的正确选择涉及工艺知识、阀门知识和执行机构知识。控制阀只有在执行机构能够承受施加在其上的静态和动态负载时才能发挥其功能。正确的选型非常重要，必须了解以下参数：

- 可用的动力源；
- 故障安全模式；
- 扭矩或推力要求（执行机构能力）；
- 控制功能；
- 经济性；
- 规模大小、模块化结构、维护性。

最广泛使用的控制阀的执行器是气动薄膜式（图 7-7）。气动执行器所需的气压可以直接来自气动控制器的输出，也可以来自将电信号转换为气压信号的信号转换器，这种转换器被称为电/气转换器，因为它们通常将 4～20mA 的电流信号转换为 20～100kPa 的气压信号。薄膜式执行器非常简单，成本低，可靠性高。许多薄膜式执行器还提供可调弹簧或其他的弹簧选择，使执行机构能适应一些特定的应用需求。由于薄膜式执行器很少有导致故障的运动部件，所以它们非常可靠。一旦出现故障，维护也很简单。

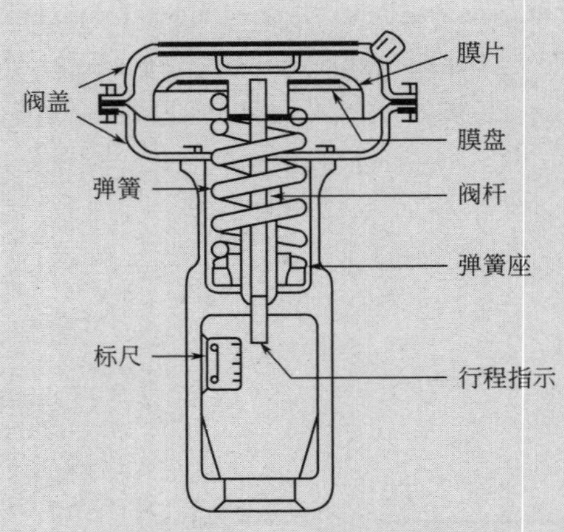

图 7-7　气动薄膜式执行器

在选择阀门时，必须考虑故障安全模式。阀门工程师必须在故障关闭阀（FC）和故障开启阀（FO）之间进行选择，这种选择是基于控制阀空气失效时过程安全的考虑。

气开式阀门是故障关闭阀（FC），当发生故障时，弹簧会关闭阀门。气开式阀门是一种故障开启阀（FO），当发生故障时，弹簧会开启阀门。

故障安全模式涉及执行机构和内阀的动作选择。执行机构和内阀通常都提供一种增加空气压力的选择，以推动阀杆向下或向上。推动阀杆下移既可以打开也可以关闭内阀。执行机构和内阀的正确组合是出于故障安全考虑，必须对阀门的工艺要求进行研究，以确定在仪表空气失效时，阀门是全开、全关，还是保持在最后的位置。图 7-8 总结了可用的阀故障安全模式。

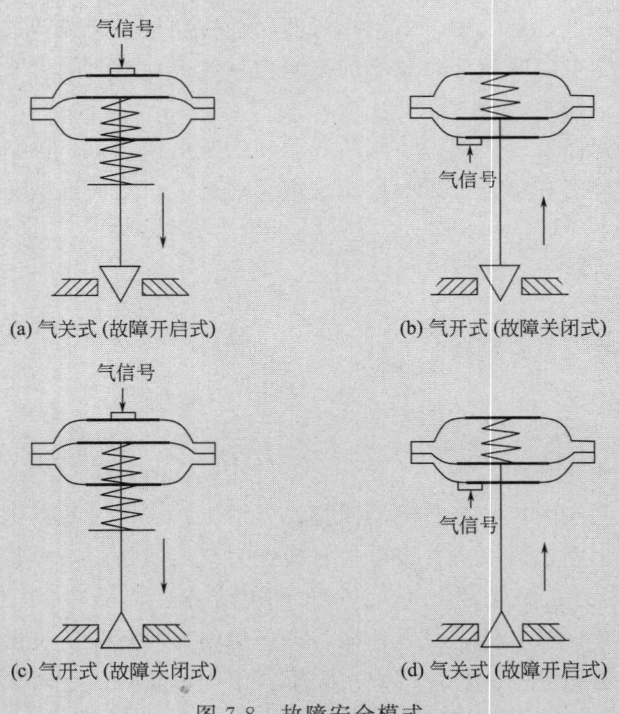

图 7-8　故障安全模式

举个例子：对于发动机自动冷却系统来说，控制阀采用故障关闭（FC）模式比采用阀门故障开启（FO）模式更危险，因为如果冷却阀门故障关闭了，发动机一定会因为没有冷却而过热；如果采用的是故障开启（图 7-9），发动机将运行在比设计条件更冷的状况下，唯一的负面影响就是效率降低。考虑到这一点，控制阀最好的选择就是故障开启模式（气关阀）。

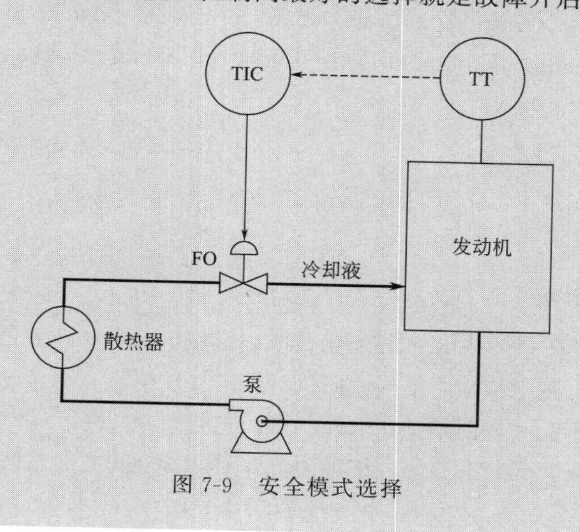

图 7-9　安全模式选择

7.5 Control Valve Positioners

Valve positioners (see Figure 7-10) are instruments that help improve control by accurately positioning a control valve actuator in response to a control signal. Positioners receive an input signal either pneumatically or electronically and provide output power, generally pneumatically, to an actuator to assure valve positioning. A positioner is a high gain proportional controller which measures the stem position, to within 0.1 mm, compares this position to a setpoint, which should be considered as the output of the main process controller, and performs correction on any resultant error signal.

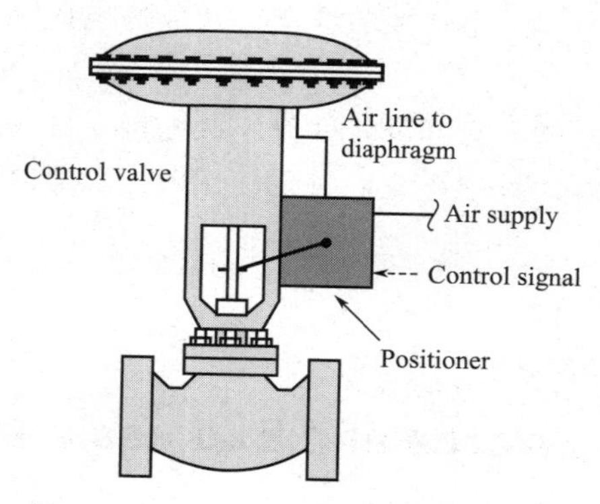

Figure 7-10　Position of the positioner

Positioners can reduce the effects of many dynamic variations, these include changes in packing friction due to dirt, corrosion, lubrication, or lack of lubrication, variations in the dynamic forces of the process, sloppy linkages (causing dead band), and nonlinearities in the valve actuators.

Remembering that a positioner becomes an intrinsic part of the full control loop very much like the secondary controller in a cascaded system, care must be exercised in their uses. On very fast loops it has been found that the use of positioners will degrade performance because the response of the positioner might not be able to keep up with the system in which it is installed.

【译文】

7.5 阀门定位器

阀门定位器（图 7-10）是一种根据控制信号准确定位执行机构位置的仪表，可以改善控制性能。定位器通过气压信号或电流输入信号，向执行机构提供输出（通常是气动方式）来确保阀门的（开度）定位。定位器是一种高增益比例控制器，它测量阀杆位置（误差控制在 0.1mm 以内），与设定值进行比较，设定值是过程控制器的输出，并对两者之间的误差信号进行校正。

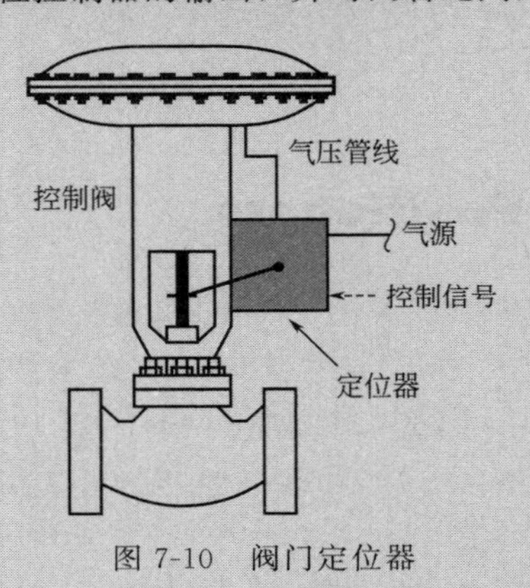

图 7-10　阀门定位器

定位器可以减少许多动态变化的影响，这些变化包括由于污垢、腐蚀、润滑或缺乏润滑而导致的填料摩擦的变化、过程动态特性的变化、松散的连接（导致死区）和阀门执行机构的非线性等。

定位器成为整个控制回路的固有部分，就像串级控制系统中的副控制器一样，在使用时必须谨慎。在一些快速的控制回路中，人们发现使用定位器会降低控制性能，因为定位器的响应可能跟不上系统的响应。

Words and Expressions　词汇和短语

1. control valve　控制阀
2. valve body　阀体
3. actuator　/'æktjueitə/　执行器
4. final element　最终执行元件

5. travel position　行程位置
6. linear　/ˈlɪniər/　adj. 线性的
7. equal percentage　等百分比
8. quick-opening　快开式
9. air-to-open　气开式
10. fail-close　故障关闭式
11. air-to-close　气关式
12. fail-open　故障开启式
13. sliding-stem　滑动阀杆
14. corrosive　/kəˈrousɪv/　adj. 腐蚀的
15. abrasive　/əˈbreɪsɪv/　adj. 磨损的
16. cavitation　/ˌkæviˈteiʃən/　n. 气穴
17. single-seated　单座
18. double-seated　双座
19. leakage　/ˈliːkɪdʒ/　n. 泄漏
20. diaphragm valve　隔膜阀
21. weir　/wɪr/　n. 堰
22. compressor　/kəmˈpresər/　n. 压缩机
23. spherical　/ˈsferɪkl/　adj. 球形的
24. cylindrical　/səˈlɪndrɪkl/　adj. 圆柱形的
25. damper　/ˈdæmpər/　n. 风门
26. throttle valve　节流阀
27. corrosion　/kəˈrouʒn/　n. 腐蚀
28. flow characteristic　流量特性
29. torque　/tɔːrk/　n. 转矩
30. thrust　/θrʌst/　n. 推力
31. pneumatic diaphragm valve　气动薄膜阀
32. I/P converter　电/气转换器
33. valve positioner　阀门定位器
34. packing friction　填料摩擦
35. lubrication　/ˌluːbriˈkeiʃn/　n. 润滑

Chapter 8

Control System Fundamentals

第 8 章

控制系统基本原理

8.1 Introduction

Control operations that involve human action to make an adjustment are called manual control systems. This type of process control is accomplished by observing a parameter, comparing it to some desired value, and initiating a control action to bring the parameter as close as possible to the desired value. For example, a human operator might have watched a level gauge and closed a valve when the level reached the setpoint. Conversely, control operations In which no human Intervention is required, such as an automatic valve actuator that responds to a level controller, are called automatic control systems.

Before we learn about automatic control systems, understand the following basic concepts and terminology.

Process variable The process variable (PV) is measured by an instrument in the field, and acts as an input to an automatic controller which takes action based on the value of it. Alternatively, the PV can be an input to a data display so that the operator can use the reading to adjust the process through manual control and supervision.

Setpoint The setpoint is a value for a process variable that is desired to be maintained.

Manipulated variable The variable to be manipulated to keep the process variable at setpoint is called the manipulated variable (MV). For instance, if we control a particular flow, we manipulate a valve to control the flow. Here, the valve position is called the MV and the measured flow becomes the PV.

Disturbance Disturbances enter or affect the process and tend to drive the process variable away from the setpoint. The control system must adjust the manipulated variable so the setpoint value of the process variable is maintained despite the disturbances.

Process A process consists of an assembly of equipment and material that is related to some manufacturing operation or sequence. Any given process can involve many dynamic variables, and it may be desirable to control all of them. In most cases, however, controlling only one variable will be sufficient to control the process to within acceptable limits.

Open-loop control loop The open-loop control loop does not compare process variables and action is taken not in response to feedback on the condition of the process variable. For example, a water valve may be opened to add cooling water to a process to prevent the process fluid from getting too hot, based on a preset time interval, regardless of the actual temperature of the process fluid. In open-loop control, the control action is not a function of the process variable. The open-loop control does not self-correct when the process variable drifts, and this may result in large deviations from the optimum value of the PV.

Closed-loop control loop The closed-loop control loop is also called a feedback control, in which the process variable is measured and compared with a setpoint, and action is taken to correct any deviation from setpoint.

【译文】

8.1 简介

需要有人直接参与的控制操作称为手动控制系统。这种类型的过程控制是通过观察参数并将其与某个期望的值进行比较，然后启动控制动作使参数尽可能接近期望的值。例如人工操作员观察液位计，并在液位达到设定值时关闭阀门。相反，不需要人工干预的控制操作，如通过自动阀门来响应液位控制器，对液位产生自动的控制动作，这种方式称为自动控制。

在我们学习自动控制系统之前，先了解以下基本概念和术语。

过程变量 过程变量（PV）由现场的仪表进行测量，并作为自动控制器的输入，自动控制器根据其值采取措施。另外，过程变量也可以作为数据显示器的输入，因此操作员可以根据读数采用手动控制和监督来调整过程。

设定值 设定值是需要保持的过程变量的目标值。

操纵变量 为了使过程变量保持在设定值，需要操纵的变量称为操纵变量（MV）。例如，如果我们可以操纵阀门来控制特定的流量，那么阀位就被称为操纵变量，流量即为过程变量。

干扰 干扰会进入或影响过程，并导致过程变量偏离设定值。控制系统必须调节操纵变量，以便即使有干扰，过程变量也能保持在设定值。

过程 过程是与制造或工业流程相关的设备和材料的总称。任何给定的过程都可能涉及许多的变量，因此可能需要控制所有这些变量。但实际上，在大多数情况下，控制一个变量就足以将过程控制在可接受的范围内。

开环控制回路 开环控制不比较过程变量，并且所采取的行动不对过程变量的条件进行反馈。例如，可以不考虑流体的实际温度，仅仅根据预先设定的时间间隔来打开水阀，添加冷却水以防止流体过热。在开环控制中，控制动作不是过程变量的函数。当过程变量偏离时，开环控制不能进行自校正，这可能导致过程变量出现较大的偏差。

闭环控制回路 闭环控制又称反馈控制。在反馈控制回路中，对过程变量进行测量，并与设定值进行比较，对偏离设定值的偏差采取纠正措施。

8.2 Process Modeling

In order to effectively control the process, we need to know how the control input we are proposing to use will affect the output of the process. If we change the input conditions we shall need to know the following:

- Will the output rise or fall?
- How much response will we get?
- How long will it take for the output to change?
- What will be the response curve or trajectory of the response?

The answers to these questions are best obtained by creating a mathematical model of the relationship between the chosen input and the output of the process in question. An understanding of a process can be obtained by developing a theoretical process model using energy balances, mass balances, and chemical and physical laws; however, this complicated and time-consuming effort can be omitted in many cases. A good approximation of process dynamics can be obtained by using simplified calculation methods, thus reducing the work required to design control systems.

In practical applications, process gain (K), time constant (T), and dead time (τ) are often used to describe the characteristics of the process, these three parameters are obtained by the following procedures:

- The control loop is opened by switching it to manual.
- Wait until the process variable (system output) is stable, and no disturbance is allowed at this time.
- A step change is applied to the manipulated variable (controller output, which is an input to the process). The step changes are usually 5% to 10% and maintained.
- According to the time-varying curve of the process variable (system output), the process parameters K, T, and τ are estimated.

Process gain (K) indicates how much a process property (output) changes per unit of input change. Process gain is determined by a number of factors, these include the physical properties of the equipment such as the vessel volumes, compressor characteristics, and connecting piping dimensions; process variables such as pressure, temperature, and fluid velocity; and various laws of physics and chemistry. The gain defines the sensitivity relating the output and input variables, it can be calculated as follows:

$$K=\frac{\Delta_{\text{Output}}}{\Delta_{\text{Input}}} \tag{8-1}$$

Time constant (T) is the amount of time required for the output variable to change 63.2% of the way from its starting point to its ultimate (terminal) value.

Dead time (τ) is also referred to as transport delay, usually caused by the transportation of material at finite speed across some distance.

A high percentage of all chemical process can be modeled by using first-order plus dead time systems, the transfer function of the process is expressed as

$$G(s)=\frac{K}{Ts+1}e^{-\tau s} \tag{8-2}$$

where

K is the process gain

T is the time constant

τ is the dead time

For example, In Figure 8-1, a 5% step change to the manipulated variable (controller output, which is an input to the process) has been applied, and process response (process output) in terms of the controlled temperature is recorded, the model parameters are

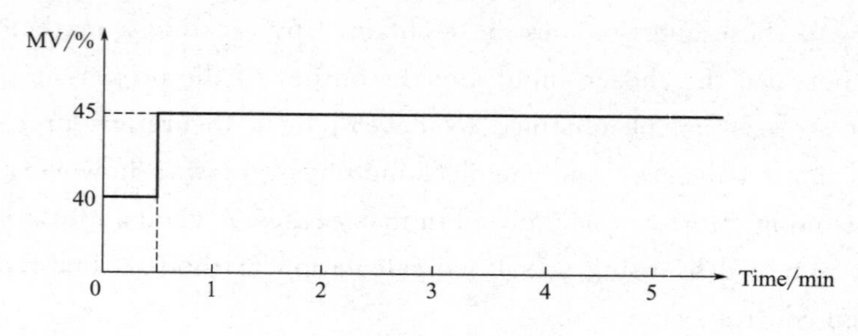

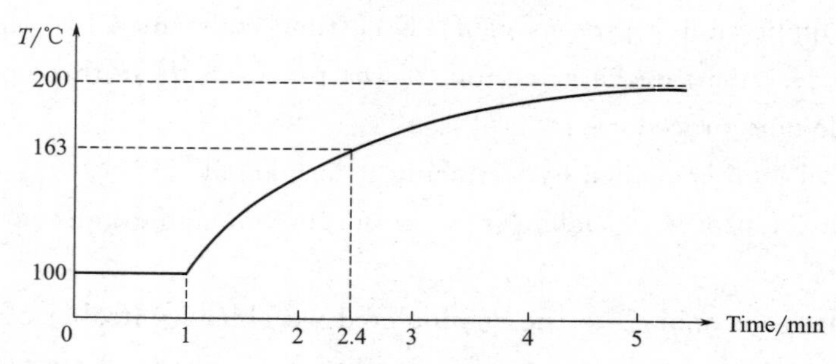

Figure 8-1 Reaction curve of a first-order plus dead time process

$$\begin{cases} K=\dfrac{\Delta_{\text{Output}}}{\Delta_{\text{Input}}}=\dfrac{100℃}{5\%}=20\ \dfrac{℃}{\%} \\ T=2.4-1=1.4\text{min} \\ \tau=1-0.5=0.5\text{min} \end{cases} \tag{8-3}$$

The presence of dead time significantly complicates the analysis and design of feedback control loops. During dead time, there is no process response and therefore no information available to initiate corrective action. It degrades the control loop's performance because it introduces an unstable behavior, which makes it difficult to achieve a satisfactory control. Special care should be taken when the dead time-to-time constant ratio exceeds 1, because the PID controller cannot handle such processes and such control structures as the Smith predictor could be required.

【译文】

8.2 过程建模

为了有效地进行控制，我们应当知道控制输入对输出的影响过程。如果改变了输入的条件，我们需要知道以下内容：

- 输出会上升还是下降？
- 会得到多大的响应？
- 输出多长时间会发生改变？
- 响应曲线或轨迹是怎样的？

通过创建所选过程输入与输出之间关系的数学模型，可以得到这些问题的答案。利用能量平衡、质量平衡以及化学和物理定律来建立理论的过程模型，可以获得对过程的理解。然而在许多情况下，可以省去这种复杂而费时的工作，使用简化的计算方法也可以很好地近似过程模型，从而减少了设计控制系统所需的工作。

在实际的应用中，常用过程增益（K）、时间常数（T）和时滞（τ）来描述过程的特性。过程参数 K、T、τ 按以下步骤确定：

• 控制回路切换到手动状态；

• 等待过程变量（系统输出）稳定，此时不允许发生扰动；

• 对操纵变量（控制器的输出，即过程对象的输入产生的输出）施加一个阶跃变化，幅度通常是 5%～10%，并一直维持；

• 记录过程变量（系统输出）随时间的变化曲线，根据曲线来估算过程参数 K、T、τ。

过程增益（K）表示过程输出的变化量相对于输入的变化量。过程增益由许多因素决定，包括设备的物理特性，如容器容积、压缩机特性和连接管道的尺寸；还与过程变量有关，如压力、温度和流体速度，以及各种物理和化学规律。过程增益定义了与输出和输入变量相关的灵敏度，表示为：

$$K=\frac{\Delta_{\mathrm{Output}}}{\Delta_{\mathrm{Input}}} \tag{8-1}$$

时间常数（T）是输出变量从其起点变化到其最终值的 63.2%所需的时间。

时滞（τ）也称为死区时间、传输延迟，通常是由于物料（或信号）在一定距离上进行传输而引起的时间延迟。

在化工过程中，有很多过程可以采用一阶加滞后系统来建模，过程的传递函数表示为：

$$G(s)=\frac{K}{Ts+1}\mathrm{e}^{-\tau s} \tag{8-2}$$

式中，K 是增益；T 是时间常数；τ 是时滞。

例如图 8-1 中，对操纵变量（由过程的输入产生的控制器输出）进行了 5%的阶跃变化，并记录了温度（过程的输出）的响应曲线。根据模型参数可求得：

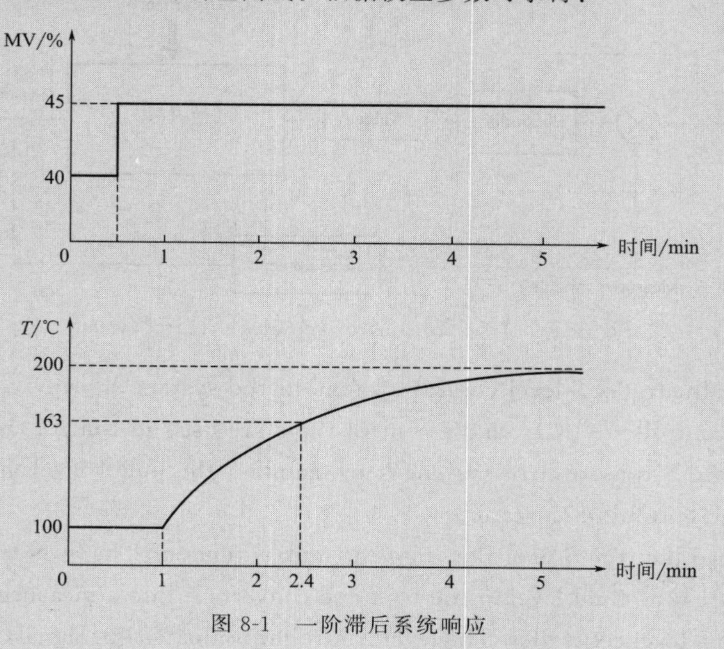

图 8-1　一阶滞后系统响应

$$\begin{cases} K=\dfrac{\Delta_{\text{Output}}}{\Delta_{\text{Input}}}=\dfrac{100℃}{5\%}=20\ \dfrac{℃}{\%} \\ T=2.4-1=1.4\text{min} \\ \tau=1-0.5=0.5\text{min} \end{cases} \tag{8-3}$$

时滞的存在，使反馈控制回路的分析和设计变得非常复杂。在时滞这段时间内，没有过程响应，因此没有可用信息来进行纠正。由于时滞导致了不稳定，使得难以实现满意的控制，从而降低了控制回路的性能。当时滞与时间常数之比超过1时应特别注意，因为PID控制器不能处理这样的过程，此时可能需要采用Smith预测器这样的控制结构。

8.3 Feedback Control System

The basic process control system consists of a closed-loop control loop as shown in Figures 8-2, also known as a feedback control system. Some elements comprise the essentials of a feedback control system: the process (the system to be controlled), the process variable (the specific quantity to be measured and controlled), the transmitter (the device used to measure the process variable and output a corresponding signal), the controller (the device that decides what to do to bring the process variable as close to setpoint as possible), the final control element (the device that directly exerts control over the process, usually a valve), and the manipulated variable (the quantity to be directly altered to effect control over the process variable).

The structure of a feedback control system can be represented by a block diagram as shown in Figure 8-2.

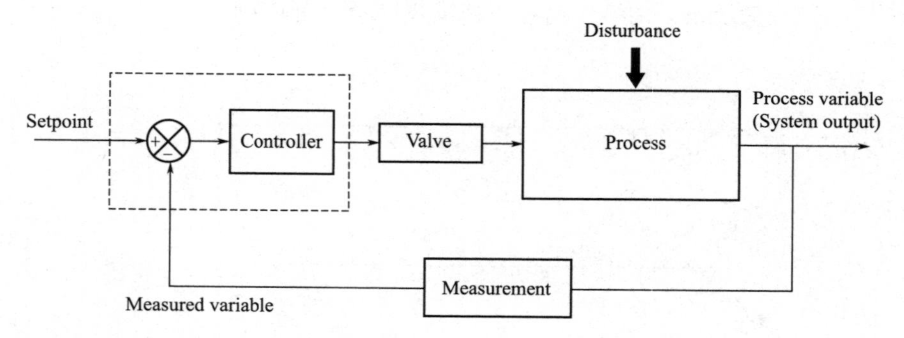

Figure 8-2 The structure of a feedback control system

Figure 8-3 illustrates a level control system. In the system shown, a level transmitter (LT), a level controller (LC), and a control valve are used to control the liquid level in a tank. The purpose of this control system is to maintain the liquid level at some prescribed height above the bottom of the tank.

It is assumed that the rate of flow into the tank is random. The level transmitter is a device that measures the fluid level in the tank and converts it into a measured variable signal, which is sent to a level controller. The level controller evaluates the signal, compares it with

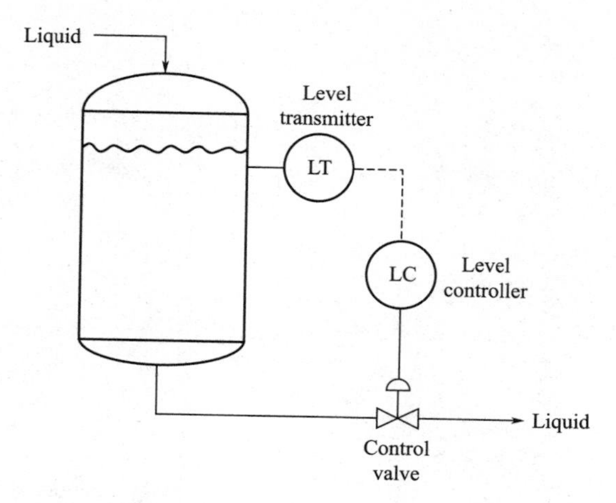

Figure 8-3 Level control system

a desired setpoint (SP), and produces a series of corrective actions that are sent to the control valve. The valve controls the flow of fluid in the outlet pipe to maintain a level in the tank.

For a well-designed feedback control system, when the disturbance input occurs, such as a step input, the output of the system will be stable again after the action of the controller, and the dynamic process is satisfactory, the system response curve is shown in figure 8-4. The performance requirements of the system are mainly reflected in three aspects: Stability, response time, steady-state error.

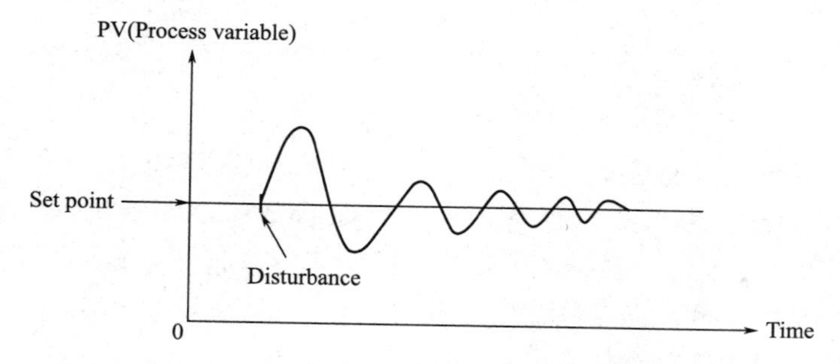

Figure 8-4 System response curve

【译文】

8.3 反馈控制系统

基本的过程控制系统由闭环控制回路组成，如图 8-2 所示，也称为反馈控制系统。反馈控制系统的要素包括：过程（要控制的系统）、过程变量（要测量和控制的变量）、变送器（用于测量过程变量并输出相应值的信号）、控制器（决定如何使过程变量尽可能接近设定值的设备）、最终控制元件（直接对过程施加控制的设备，通常是阀门）以及操纵变量（为实现对过程变量的控制而直接改变的量）。

反馈控制系统的结构用方框图表示如图 8-2 所示。

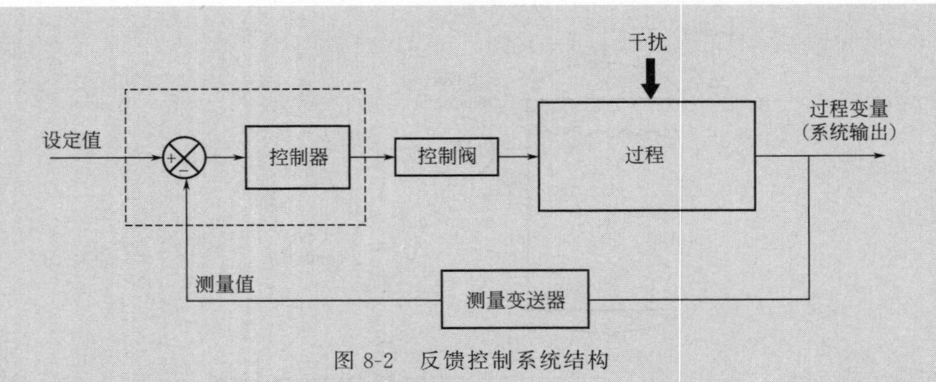

图 8-2 反馈控制系统结构

图 8-3 所示为一个液位控制系统。在所示的系统中，液位变送器（LT）、液位控制器（LC）和控制阀共同用于控制水箱中的液位。这个控制系统的目的是保持液位在规定的高度。

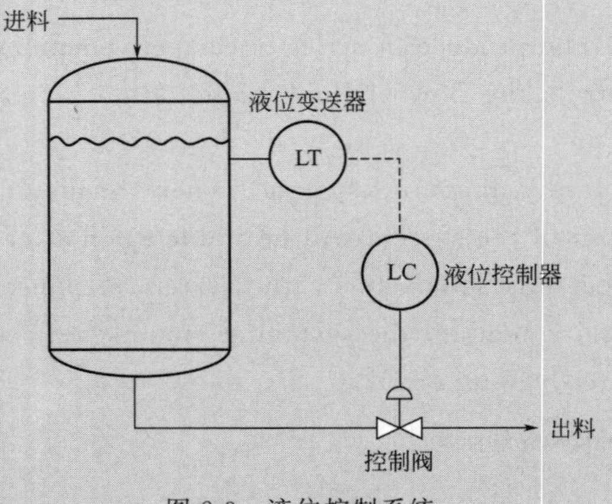

图 8-3 液位控制系统

假设流入水箱的流量是随机变化的，液位变送器是测量罐中液位并将其转换为测量变量信号的设备，该信号被发送到液位控制器。液位控制器评估该液位信号并将其与设定值（SP）进行比较，产生一系列控制纠正措施。这些纠正措施将被发送到控制阀，通过控制阀调节出口管道中的流体流量，以保持罐中的液位。

对于一个设计合理的反馈控制系统，当发生干扰输入（例如阶跃输入）时，在控制器动作后系统输出将再次稳定，并且动态过程令人满意，如图 8-4 所示的系统响应曲线。系统的性能要求主要体现在稳定性、响应时间、稳态误差三个方面。

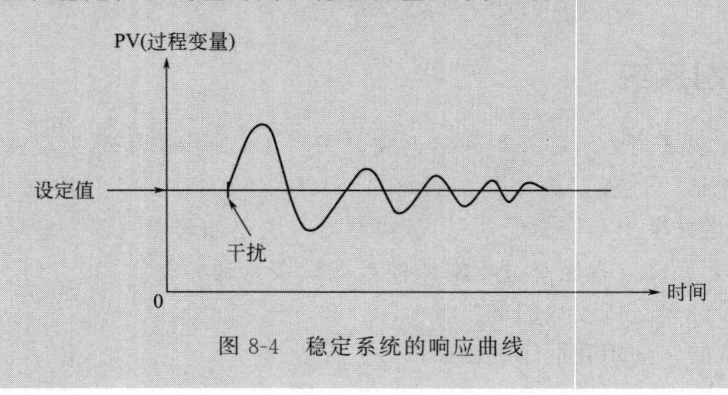

图 8-4 稳定系统的响应曲线

8.4 Tuning PID Controllers

The analysis of the transition process of the control system is the basis for measuring the performance of the control system. The transient process refers to the process in which the control system is disturbed to reach a new state from the original steady-state. The performance of a control system depends on the following:

- The quality of the measuring and control devices
- The effect of process upsets
- The control stability as manifested in the ability of the measured variable to return to its setpoint after a disturbance, this ability is dependent on the correct controller PID settings, which is accomplished through good tuning.

Generally, the transition process of a well-designed control system has a decay ratio of one-quartera, as shown in Figure 8-5.

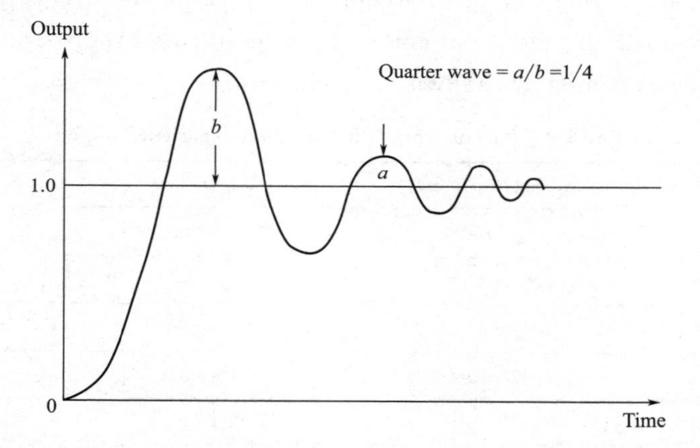

Figure 8-5 Process response curve for one-quarter decay ratio

Tuning means finding the ideal combination of P, I, and D to provide the optimum performance for the loop under operating conditions. Keep in mind that "ideal control" must be determined for a specific application.

There are two basic methods for manual tuning: open loop and closed loop. Open loop tuning may be used to tune loops that have long delays such as temperature loops, and closed loop tuning may be used to tune fast loops such as flow, pressure, and level loops. The Ziegler-Nichols closed loop method is the most commonly used.

The Ziegler-Nichols closed loop method consists of the following steps:

- Putting the process on auto control using "P only" mode (set I and D to minimum)
- Moving the controller set point 10 percent and holding until PV begins to move
- Returning the setpoint to its original value
- Adjusting the proportional band (PB) from large to small until a stable equal continuous cycle is obtained. The critical PB at this time is recorded as δ_k
- Measuring period of cycle T_k
- According to the obtained δk and T_k, set the P, I, D parameters of the controller

according to the Table 8-1

- Testing and fine tuning, if required.

Table 8-1 Parameters of the Ziegler-Nichols closed loop method

Control mode	proportional band	integral time	derivative time
P	$2\delta_k$	—	—
PI	$2.2\delta_k$	$0.85T_k$	—
PD	$1.8\delta_k$	—	$0.1T_k$
PID	$1.7\delta_k$	$0.5T_k$	$0.125T_k$

Besides Ziegler-Nichols closed loop method, there is a method called Based-on-Experience Tuning. When using this method, known values of P, I, and D are entered. This is a rough way of doing controller tuning, and it does not generally work from the first trial. To make it work, repeated "fine tuning" is required: tweaking the PID settings until acceptable settings are obtained through trial-and-error adjustments. Approximate typical settings for based-on-experience tuning are shown in Table 8-2.

Table 8-2 Parameters of the based-on-experience tuning

Loop type	proportional band	integral time	derivative time
Flow	40%～150%	0.3～1min	—
Level	20%～80%	—	—
Temperature	20%～60%	3～10min	0.5～3min
Pressure	30%～70%	0.4～3min	—

【译文】

8.4 PID参数整定

对控制系统过渡过程的分析是衡量控制系统性能的基础。过渡过程是指控制系统的输出从原始稳态进入新的稳态的过程。控制系统的性能取决于以下几点：

- 测量和控制仪表的质量；
- 过程扰动的影响；
- 控制的稳定性，表现为被测变量在受到干扰后能够回到其设定值，这种能力取决于正确的控制器PID参数的设置，需要通过参数整定（调优）来实现。

通常，设计良好的控制系统的过渡过程的衰减比为4∶1，如图8-5所示。

参数整定指的是找到P、I和D的理想组合，以便在操作条件下为控制回路提供最佳性能，特定的控制系统有特定的整定参数。

手动参数整定有两种基本方法：开环整定和闭环整定。开环整定可用于整定具有较大滞后的回路，如温度回路；闭环整定用于快速响应的回路，如流量、压力和液位控制回路。Ziegler-Nichols闭环方法是最常用的。

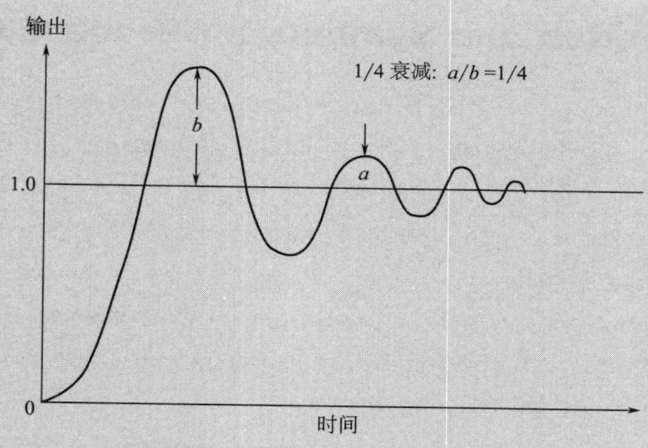

图 8-5 4：1 衰减过渡过程

Ziegler-Nichols 闭环整定方法包括以下步骤。

- 在纯比例状态下，将过程投入自动控制状态（将积分和微分设置为最小）；
- 将设定值调整 10%并保持，直到 PV 开始移动；
- 将设定值切回到初始值；
- 从大到小调整比例度（PB），直到获得稳定的等幅振荡（获得临界比例度 δ_k）；
- 测量振荡周期 T_k；
- 根据获得的临界比例度 δ_k 和振荡周期 T_k，按照表 8-1 设置控制器的 P、I、D 参数；
- 如果需要，进行测试和微调。

表 8-1 Ziegler-Nichols 闭环方法的参数设置

控制模式	比例度	积分时间	微分时间
P	$2\delta_k$	—	—
PI	$2.2\delta_k$	$0.85T_k$	—
PD	$1.8\delta_k$	—	$0.1T_k$
PID	$1.7\delta_k$	$0.5T_k$	$0.125T_k$

除了 Ziegler-Nichols 闭环整定外，还有一种基于经验的整定方法。使用此方法时，将已知的 P、I 和 D 经验参数值直接输入。这是一种进行控制器整定的粗略方法，通常刚开始无法正常工作，需要重复地“微调”来获得最优值。基于经验的整定采用表 8-2 的近似典型设置。

表 8-2 基于经验法的参数设置

回路类型	比例度/%	积分时间/min	微分时间/min
流量	40～150	0.3～1	—
液位	20～80	—	—
温度	20～60	3～10	0.5～3
压力	30～70	0.4～3	—

8.5 Identification and Symbols for Process Control

When drawing Piping and Instrumentation Diagrams (referred to as P&ID), a unified method is needed to describe and identify all classes of instruments, instrumentation systems, and functions used for measurement, monitoring, and control. It is done by presenting a designation system of graphic symbols and identification letter codes.

(1) Graphic symbol system

The graphic symbol system shall be used to depict instrumentation in text and in sketches and drawings. When used with identification letters and numbers, it shall identify the functionality of each device and function shown.

Table 8-3 shows discrete instruments in non-microprocessor systems, such as single-case transmitters, controllers, indicators, or recorders.

Table 8-3 Discrete devices

Graphic Symbol	Application
(circle)	Field or locally mounted Normally accessible to an operator
(circle with single solid horizontal line)	Central or main control room Front of main panel mounted
(circle with double solid horizontal line)	Field or local control panel Front of secondary or local panel mounted
(circle with single dashed horizontal line)	Central or main control room Rear of main panel mounted
(circle with double dashed horizontal line)	Field or local control panel Rear of secondary or local panel or cabinet mounted

Table 8-4 shows shared or distributed software analog instruments in distributed control or programmable logic control systems.

Table 8-4 Shared continuous devices

Graphic Symbol	Application
(circle in square)	Dedicated single function device Field or locally mounted Normally accessible to an operator at device
(circle in square with single solid horizontal line)	Central or main console Visible on video display Normally accessible to an operator at console
(circle in square with double solid horizontal line)	Secondary or local console Field or local control panel Visible on video display Normally accessible to an operator at console

Continue

Graphic Symbol	Application
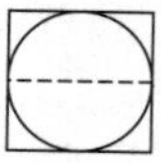	Central or main console Not visible on video display Not normally accessible to an operator at console
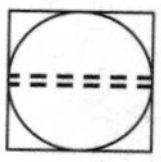	Field or local control panel Not visible on video display Not normally accessible to an operator at console

Table 8-5 shows software instruments in microprocessor-based control systems similar or equal to a distributed control or programmable logic control systems.

Table 8-5 Shared logic control devices

Graphic Symbol	Application
	Field or locally mounted Not panel or cabinet mounted Normally accessible to an operator at device
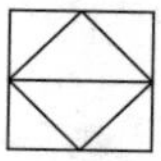	Central or main console Visible on video display Normally accessible to an operator at console
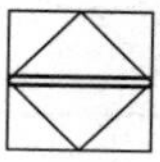	Secondary or local console Field or local control panel Visible on video display Accessible to an operator at console
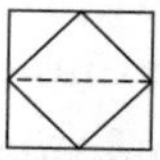	Central or main console Not visible on video display Not normally accessible to an operator at console
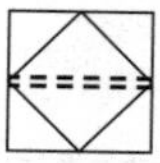	Secondary or local console Field or local control panel Not visible on video display Not normally accessible to an operator at console

Table 8-6 shows shared or distributed on-off software instruments in a computer-based control system.

Table 8-6 Computer devices

Graphic Symbol	Application
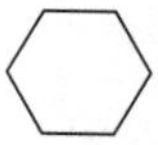	Undefined location Undefined visibility Undefined accessibility
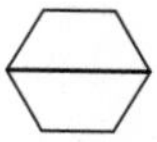	Central or main computer Visible on video display Accessible to an operator at console or computer terminal
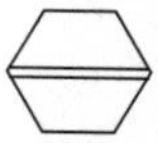	Secondary or local computer Visible on video display Accessible to an operator at console or computer terminal

Continue

Graphic Symbol	Application
	Central or main computer Not visible on video display Not accessible to an operator at console or computer terminal
	Secondary or local computer Not visible on video display Not accessible to an operator at console or computer terminal

Figure 8-7 shows the valve bodies of some common control valves.

Table 8-7 Valve body symbols

Valve name	Symbol	Valve name	Symbol
General symbol		Three-way	
Angle		Four-way	
Butterfly		Globe	
Rotary valve		Diaphragm	

(2) Instrument line symbols

Instrument line symbols (see Table 8-8), contains lines used to represent process connections and the measurement and control signals that connect instruments and functions to the process and to each other.

Table 8-8 Instrument line symbols

Symbol	Application
	Instrument supply or connection to process
	Pneumatic signal
or	Electronic signal
	Hydraulic signal
	Guided electromagnetic or sonic signal
	Unguided electromagnetic or sonic signal
	Communication link or system bus
	Mechanical link

(3) Identification letter tables

Identification letter tables provide the alphabetic building blocks of the instrument and function identification system in a concise, easily referenced manner. The typical tag number consists of two parts: a functional identification and a loop number (e. g. , LIC-101). The functional identification consists of a first letter and one or more succeeding letters. The first letter designate the measured variable, for example, P for pressure, T for temperature, etc. The succeeding letters identify the functions performed, for example, I for indicate, T for transmit, C for control, etc. See Table 8-9 for details. For example, a temperature indicating controller is identified as TIC, a flow transmitter as FT; a temperature recorder as TR, a level controller as LC, and so on.

Table 8-9 Identification letters table

	First letters		Succeeding letters
	Measured variable	Modifier	Function
A	Analysis		Alarm
C	Conductivity		Control
D	Density	Differential	
E	Voltage		Sensor
F	Flow Rate	Ratio	
I	Current		Indicate
L	Level		Light
M	Humidity		
P	Pressure		Point
Q	Quantity	Integrate, totalize	
R	Radiation		Record
S	Speed, frequency	Safety	Switch
T	Temperature		Transmit
V	Vibration, mechanical analysis		Valve, damper, louver
W	Weight, force		Well
Y	Event, state or presence		Relay, compute, convert
Z	Position		Driver, actuator

【译文】

8.5 过程控制的标识和符号

在绘制管道仪表流程图（称为 P&ID）时，需要一种统一的方法来描述和标识用于测量、监视和控制的所有类别的仪器、仪表系统和功能，通常采用图形符号和字母代号来完成。

(1) 图形符号

图形符号用以在文本、草图和绘图中描绘仪表。当与字母代号和数字一起使用时，它可以表示所显示的每个设备的功能。

表 8-3 列出了在常规仪表系统中的离散设备，如单体的变送器、控制器、指示器或记录仪。

表 8-3 离散设备

图形符号	应用
	就地安装,现场操作
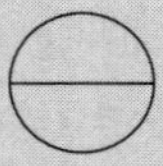	控制室集中盘面安装
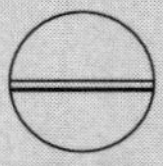	就地仪表盘面安装
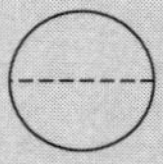	控制室盘后安装
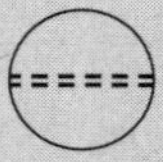	就地仪表盘后安装

表 8-4 列出了分布式控制或可编程逻辑控制系统中的模拟仪表。

表 8-4 分布式控制共享设备

图形符号	应用
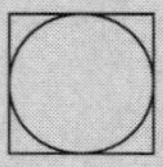	专用单功能设备,现场安装 操作员通常可以访问
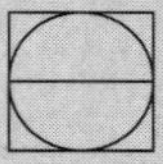	主控制台,显示器上可见 通常由控制台的操作员访问
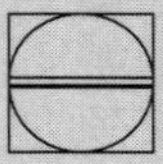	辅助控制台,显示器上可见 通常由控制台的操作员访问
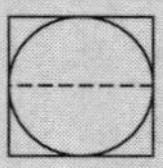	主控制台,显示器上不可见 操作员无法在控制台上访问
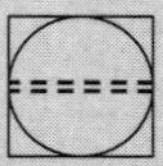	现场控制面板,显示器上不可见 操作员通常无法在控制台上访问

表 8-5 列出了基于微处理器的控制系统中的软件仪器，它们与分布式控制或可编程逻辑控制系统类似。

表 8-5 共享逻辑控制设备

图形符号	应用
	现场或就地安装 操作员可以访问
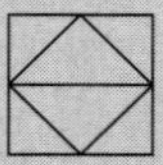	主控制台,显示器上可见 由控制台的操作员访问

续表

图形符号	应用
	辅助控制台，显示器上可见 操作员可通过控制台访问
	主控制台，显示器上不可见 操作员无法在控制台上访问
	辅助控制台，显示器上不可见 操作员无法在控制台上访问

表 8-6 列出了计算机控制系统中的共享或分布式开关软件工具。

表 8-6 计算机设备

图形符号	应用
	未定义的位置，可见度不确定 不确定的可访问性
	主计算机 在视频显示器上可见 操作员可通过控制台或计算机终端访问
	辅助计算机 在视频显示器上可见 操作员可通过控制台或计算机终端访问
	主计算机 在视频显示器上不可见 操作员无法在控制台或计算机终端上访问
	辅助计算机 在视频显示器上不可见 操作员无法在控制台或计算机终端上访问

表 8-7 是部分常用控制阀的阀体的表示形式。

表 8-7 阀门符号

阀门名称	符号	阀门名称	符号
通用符号		三通阀	
角阀		四通阀	
蝶阀		截止阀	
旋转阀		隔膜阀	

(2) 仪表连接线符号

表 8-8 是仪表连接线符号。仪表连接线符号包括用于表示过程连接的线条，以及将仪表和功能与过程相互连接的测量和控制信号线。

表 8-8　仪表连接线符号

符号	应用
	仪表或过程连接线
	气信号连接线
或	电信号连接线
	液压信号连接线
	导向电磁或声波信号
	非导向电磁或声波信号
	通信链接或系统总线
	机械连接

(3) 字母代号表

字母代号表以简明易懂的方式提供了仪表和功能识别系统的字母组成方式。典型的位号由两部分组成：功能识别号和回路号（例如 LIC-101）。功能标识由首字母和一个或多个后续字母组成。首字母表示被测变量，例如 P 表示压力，T 表示温度等；后续字母表示执行的功能，例如 I 表示指示，T 表示变送，C 表示控制等。表 8-9 进行了详细说明，例如温度指示控制器标识为 TIC；流量变送器标识为 FT，温度记录器为 TR，液位控制器为 LC 等等。

表 8-9　字母代号表

	首字母		后续字母
	被测变量	修饰词	功能
A	分析		报警
C	电导率		控制
D	密度	差	
E	电压		检测元件
F	流量	比率(比值)	
I	电流		指示
L	物位		灯
M	湿度		
P	压力		连接或检测点
Q	数量	积分，累计	
R	放射性		记录
S	速度，频率	安全	转换开关
T	温度		变送
V	振动，机械分析		阀门，风门，百叶窗
W	重量，力		套管
Y	事件，状态		继电器，计算器，转换器
Z	位置		驱动，执行

Words and Expressions 词汇和短语

1. manual control 人工控制
2. automatic control 自动控制
3. process variable 过程变量
4. manipulated variable 操纵变量
5. disturbance /dɪ'stɜːrbəns/ n. 干扰
6. process /'prəuses/ n. 过程
7. open-loop control 开环控制
8. closed-loop control 闭环控制
9. supervision /ˌsjuːpə'viʒən/ n. 监督
10. response curve 响应曲线
11. mathematical model 数学模型
12. energy balance 能量平衡
13. mechanism /'mekənɪzəm/ n. 机制
14. stability /stə'bɪləti/ n. 稳定性
15. response time 响应时间
16. steady-state error 稳态误差
17. tuning /'tjʊnɪŋ/ 整定
18. transition process 过渡过程
19. trial-and-error n. 反复试验法
20. Piping and Instrumentation Diagrams 管道与仪表流程图（P&ID）
21. graphic symbol 图形符号
22. identification letter 识别字母

Chapter 9

Complex Control Systems

第9章

复杂控制系统

9.1 Cascade Control Systems

A cascade control system consists of two feedback control loops, one nested inside the other. The output signal of one controller is connected to the setpoint of another controller, each controller perceives different aspects of the same process. The first controller (called the primary controller) essentially gives orders to the second controller (called the secondary controller) via a remote setpoint signal. When two controllers are cascaded, each will have its own measurement input or PV, but only the primary controller can have an independent SP and only the secondary, or the most downstream, controller has an output to the control valve. The structure of the cascade control system is shown in Figure 9-1.

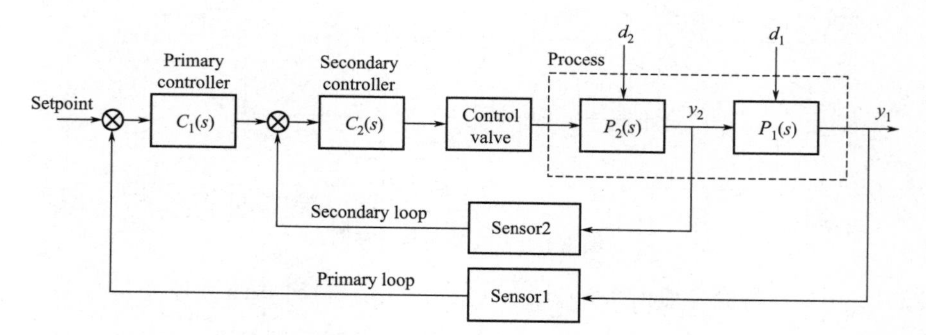

Figure 9-1 Cascade control system

The primary controller (C_1) generates the setpoint for the secondary controller (C_2), the secondary measurement (y_2) is an intermediate process variable that can be used to achieve more effective control of the primary process variable (y_1). The secondary controlled variable (y_2) affects the primary process (P_1) and therefore it also affects the primary controlled variable (y_1).

Cascade control is advantageous on applications where the primary process has a large dead time and the time delays in the secondary process are smaller. To maintain stability, the secondary loop must be much faster than the primary loop, and the secondary loop must receive the maximum disturbances (instead of—and before they affect—the primary loop), this is because with the cascade configuration, the correction of the inner disturbance d_2 occurs as soon as the secondary sensor detects that upset.

The principal advantages of cascade control are the following:

- Disturbances occurring in the secondary loop are corrected by the secondary controller before they can affect the primary variable.
- The secondary controller can significantly reduce phase lag in the secondary loop, thereby improving the speed or response of the primary loop.
- Gain variations due to nonlinearity in the process or actuator in the secondary loop are corrected within that loop.

【译文】

9.1 串级控制系统

串级控制系统由两个反馈控制回路组成，一个嵌套在另一个的内部。一个控制器的输出信号连接到另一个控制器的设定值，每个控制器感知同一过程的不同方面。第一个控制器（称为主控制器）通过外部设定值信号下达指令给第二个控制器（称为副控制器）。当两个控制器串联时，每个控制器有自己的测量值，但是只有主控制器可以具有独立的设定值，只有副控制器可以直接作用于控制阀。串级控制系统的结构如图 9-1 所示。

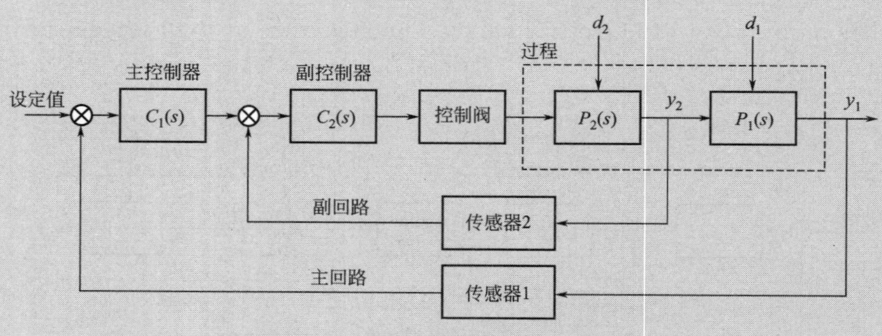

图 9-1 串级控制系统结构

副控制器（C_2）的设定值由主控制器（C_1）产生，副回路的被控变量（y_2）是中间过程变量，可用于实现对主被控变量（y_1）的更有效控制。副被控变量（y_2）影响主过程（P_1），因此也影响了主被控变量（y_1）。

串级控制适用于主回路过程滞后时间较长、副回路过程响应迅速的控制应用。为了保持稳定性，副回路必须比主回路快得多，并且副回路必须承受最大的干扰（在此干扰影响到主回路之前）。这是因为在串级控制系统结构中，一旦副传感器检测到干扰，就会立即产生对副回路干扰 d_2 的校正控制。

串级控制的主要优点如下：

- 在副回路中发生的干扰，会在影响主变量之前被副控制器校正；
- 副控制器可以大大减小副回路中的相位滞后，从而提高主控制回路的控制速度或响应速度；
- 由于控制过程或副回路中执行器的非线性而导致的增益变化，可以在副回路内得到有效校正。

9.2 Ratio Control Systems

The ratio control system refers to the control system which realizes that two or more variables accord with a certain proportion. Ratio control is applied almost exclusively to fluids. In chemical refining and other industrial production processes, it is often necessary to maintain a certain ratio of two or more fluids. If the ratio is out of balance, an accident will occur. Many industrial processes also require the precise mixing of two or more ingredients to produce a desired product. Not only do these ingredients need to be mixed in proper proportion, but it is usually desirable to have precise control over the total flow rate as well.

For the general problem of ratio control, an automated approach is to install a flow transmitter on one line and a flow control loop on the other. The signal coming from the uncontrolled flow transmitter becomes the setpoint for the flow control loop.

In ratio control system (see Figure 9-2), the controlled flow follows in proportion to a second variable known as the "wild flow". The flow transmitter on the uncontrolled line measures the flow rate of wild flow, sending a flow rate signal to the flow controller (FIC) which acts to match flow rates. If the calibrations of each flow transmitter are precisely equal to one another, the ratio of controlled flow to wild flow will be 1 : 1 (equal). The only purpose of a ratio control system is to match the two flows to a ratio, regardless of total flow rate.

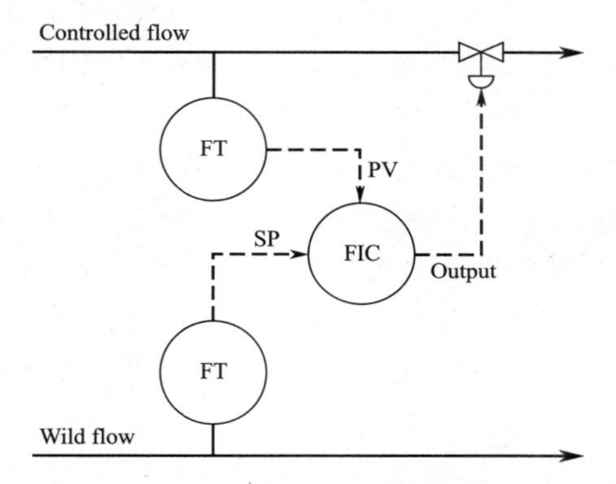

Figure 9-2 Example of a ratio control system

We may incorporate convenient ratio adjustment into this system by adding another component called a ratio station to the control scheme (see Figure 9-3). This component (or computer function) takes the flow signal from the wild flow transmitter and multiplies it by some constant value (k) before sending the signal to the controlled flow controller as a setpoint. With identical flow range calibrations in both flow transmitters, this multiplying constant k directly determines the ratio (i. e. the ratio will be 1 : 1 when $k = 1$; the ratio will be 2 : 1 when $k = 2$, etc.).

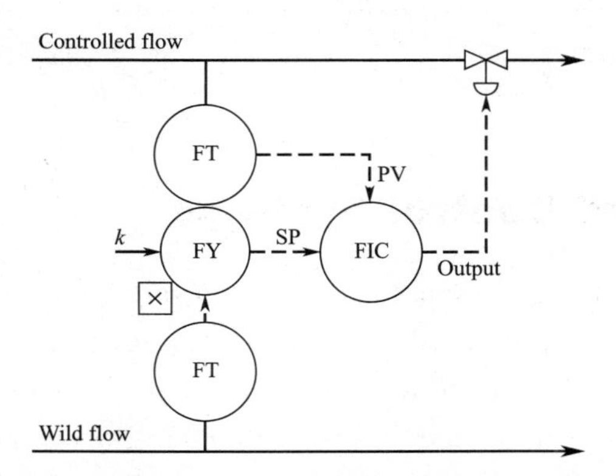

Figure 9-3 Example of a system with ratio adjustment

【译文】

9.2 比值控制系统

比值控制系统是指实现两个或两个以上变量符合一定比例关系的控制系统。比值控制几乎完全应用于流体。在化工炼油及其他工业生产过程中，工艺上常需要将两种或两种以上的流体保持一定的比例关系，因为比例一旦失调，将会造成事故。许多工业过程也需要两种或多种原料的精确混合才能生产出所需的产品。这些成分不仅需要以适当的比例混合，而且通常还需要精确控制总流量。

对于一般的比值控制问题，自动化的方法是在一条管线上安装流量变送器，在另一条管线上安装一个流量控制回路，来自不受控管线上的流量变送器的信号成为流量控制回路的设定值。

在比值控制中（图 9-2），被控流量与称为“野流”的不受控流量成比例。不受控制管路上的流量变送器测量“野流”值，将信号送到流量控制器（FIC）以匹配流量比值。如果两个流量变送器的标定彼此相等，则被控流量与“野流”之比将为 1∶1（相等）。比值控制系统的唯一目的是使这两种流量的比值恒定，而无论它们的总流量如何。

通过在控制方案中添加一个称为比值器的组件，我们可以在系统中进行便捷的比值调整（图 9-3）。该组件（或计算机功能）从“野流”变送器获取流量信号，并将其乘以某个常数 k，然后再将信号发送到被控流量控制器作为其设定值。在两个流量变送器中使用相同的流量范围校准时，这个乘法常系数 k 就直接确定了比值（即当 $k=1$ 时比值为 1∶1；在 $k=2$ 时比值为 2∶1，依此类推）。

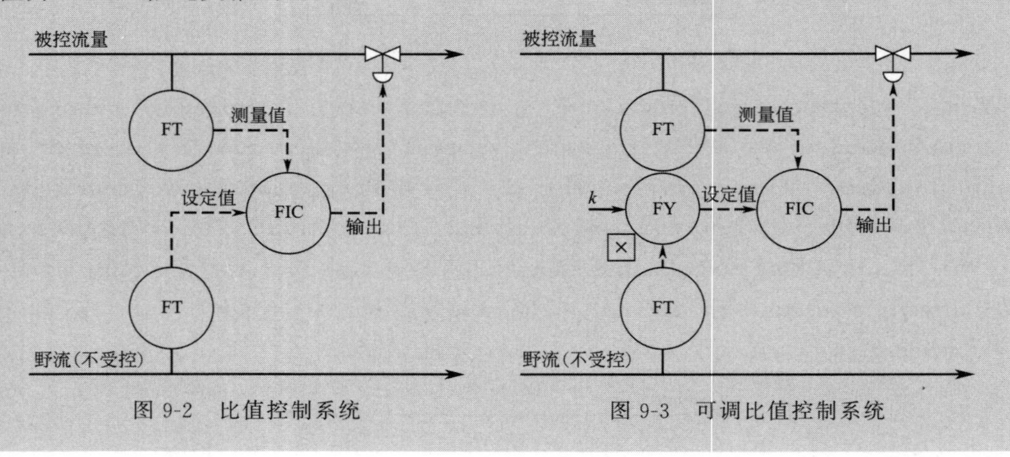

图 9-2 比值控制系统　　图 9-3 可调比值控制系统

9.3 Feedforward Control Systems

If large and random changes occur to the process variable in the feedback system, simple feedback action becomes very ineffective in correcting these large deviations. These deviations usually drive the process beyond its operating range, and the feedback controller has little opportunity to make accurate and fast corrections to the error, the result of this is that the performance of the process becomes unacceptable.

Feedforward control is used to detect and correct these disturbances before they have a chance to enter and upset the closed or feedback loop characteristics. The difference between feedforward and feedback control can be considered as:

- Feedforward is primarily designed and used to prevent errors (process disturbances) entering or disturbing a control loop within a process system. These errors can be foreseen and corrected by feedforward control, prior to them upsetting the control loop parameters.
- Feedback is used to correct errors, caused by process disturbances, that are detected within a closed loop control system.

The feedforward itself is open loop, so the disturbance variable is not controlled. Feedforward control is generally not used alone, it is generally used in conjunction with feedback control to trim the feedforward model. Figure 9-4 is an example of energy-balance feedforward-feedback control system.

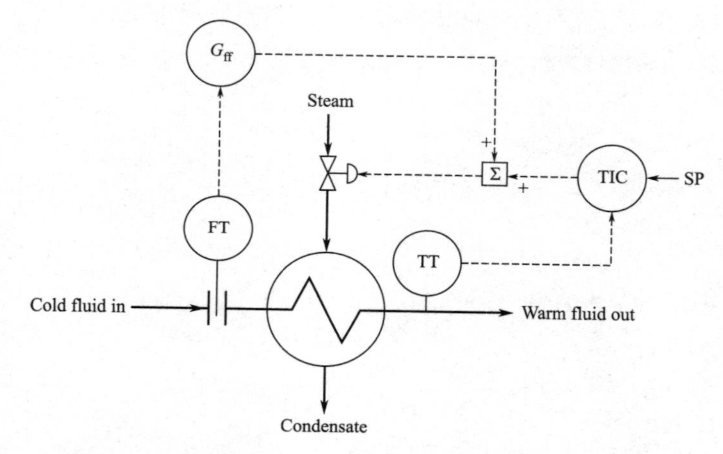

Figure 9-4 Example of a feedforward-feedback control system

The cold fluid enters the heat exchanger and becomes warm fluid output after being heated by the steam. In this control system, the controlled variable is the temperature of the warm fluid detected by a temperature transmitter. The temperature controller TIC adjusts the opening degree of the steam valve to ensure the stability of the temperature. This is a common method of feedback control.

If the flow of the cold fluid changes drastically, it will be the main source of disturbance that affects the temperature of the warm fluid. If only feedback control is used, the effect of the control is not very good due to the large lag of the heat exchanger. Therefore, we can design a feedforward-feedback control system as shown in Figure 9-4.

When the flow rate of the cold fluid changes greatly, on the one hand, the feedforward controller G_{ff} immediately acts according to some control law, changing the opening of the steam valve to compensate for the temperature of the warm fluid. On the other hand, the temperature change that can not be completely eliminated by the feedforward control is adjusted by the temperature controller TIC according to the conventional feedback control.

【译文】

9.3 前馈控制系统

如果反馈系统中的过程变量发生了较大且随机的变化，简单的反馈动作在纠正这些较大的偏差时将变得非常无效。这些偏差通常会使过程超出其工作范围，反馈控制器几乎没有机会对误差进行准确和快速的修正，其结果是过程的性能变得不可接受。

在这些干扰进入并破坏闭环或反馈回路特性之前，前馈控制可检测并纠正这些干扰。前馈控制和反馈控制的区别如下。

· 前馈主要用于防止错误（过程干扰）进入或干扰过程系统内的控制回路，这些误差可以通过前馈控制在它们扰乱控制回路参数之前进行预见和纠正。

· 反馈用于纠正闭环控制系统中检测到的过程干扰引起的偏差。

前馈本身是开环的，因此扰动变量不受控制。前馈控制通常不单独使用，它通常与反馈控制一起使用，以修正前馈模型。图 9-4 是一个采用前馈-反馈控制的能量平衡系统的实例。

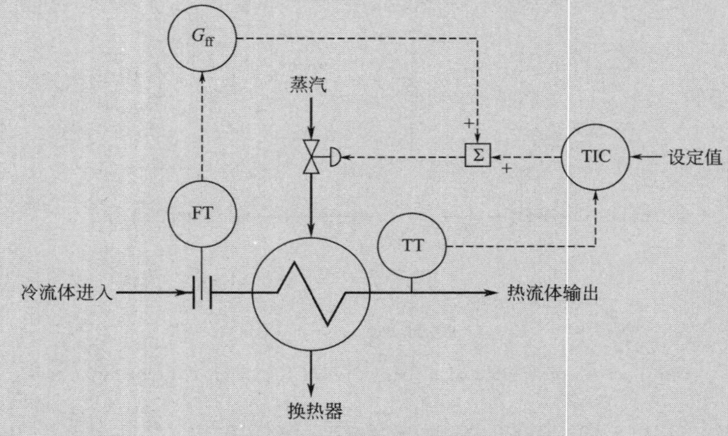

图 9-4 前馈-反馈控制系统

冷流体进入换热器，经蒸汽加热后成为热流体输出。在该控制系统中，被控变量是温度变送器检测到的热流体温度。温度控制器 TIC 调节蒸汽阀的开度，保证温度的稳定。这是一种常见的反馈控制方法。

如果冷流体的流量发生剧烈变化，将成为影响热流体温度的主要干扰源。如果只采用反馈控制，由于换热器的滞后较大，控制效果不是很好。因此，我们可以设计一个前馈-反馈控制系统，如图 9-4 所示。

当冷流体的流量变化较大时，一方面前馈控制器 G_{ff} 根据某种控制规律立即动作，通过改变蒸汽阀的开度来补偿热流体的温度；另一方面，前馈控制不能完全消除的温度变化，再由温度控制器 TIC 根据传统的反馈控制进行调节。

9.4 Selective Control Systems

Selective control is also called override control. In selective control applications, the signal selectors choose the lowest, highest, or median signal from among two or more control signals. Such selectors can be analog instruments or software function blocks in DCS control systems.

Selective control usually keep some process variable from reaching an unsafe condition, and therefore the interlock trip points are not reached. Thus, selective control keeps the equipment running although perhaps at a suboptimal level. One controller can take command of a manipulated variable away from another controller when otherwise the process would exceed some process or equipment limit or constraint.

An example is shown in Figure 9-5, where a water pump is driven by a variable-speed electric motor to draw water from a well and provide constant water pressure to customers.

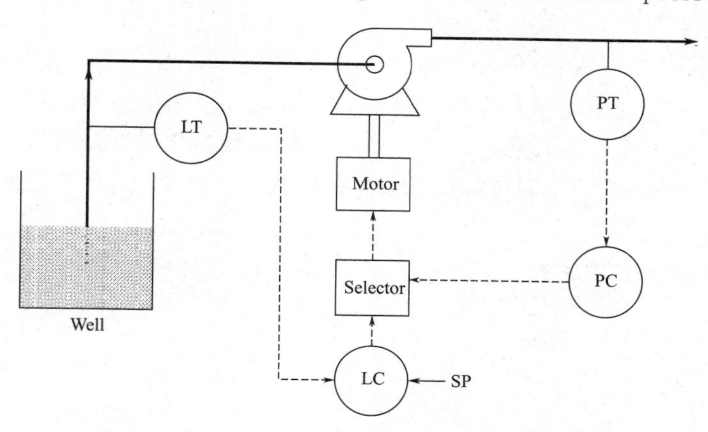

Figure 9-5 Example of a selective control system

In this system, the speed of the motor is generally adjusted according to the customer's water consumption, that is, the motor is controlled by the pressure controller PC.

However, if the customer's water demand is high, the water level in the well will be very low, and even the pump will not be able to pump water. If the pump is operated for too long without water passing through, the seal will be damaged, which is not allowed. One way to solve this problem is to install a liquid level switch to detect the water level. If the water level is too low, turn off the motor that drives the pump. But this is only a rough solution, because it protects the pump from damage, but at the cost of completely shutting off water to customers.

A better solution to the problem would be to have the pump merely slow down as the well water level approaches a low-level condition. This way at least the pump could be kept running (and some amount of pressure maintained), decreasing demand on the well while maintaining curtailed service to customers and still protecting the pump from dry-running. This would be termed a selective control system.

We may create just such a control strategy by connecting the level transmitter to the level controller LC, and using a selector or function block to select the lower value between the pressure and level controllers. The level controller's setpoint will be set at some low level above the acceptable limit for continuous pump operation. If ever the well's water level goes below this setpoint, the level controller will command the pump to slow down, even if the pressure controller is calling for a higher speed. The level controller will have overridden the pressure controller, prioritizing pump longevity over customer demand.

【译文】

9.4 选择性控制系统

选择性控制也称为超驰控制。在选择性控制应用中，信号选择器从两个或多个信号中选择最低、最高或中间信号。这样的选择器既可以是模拟量仪表，也可以是 DCS 控制系统中的软件功能块。

选择性控制可以阻止某些过程变量达到不安全状态，因此不会导致控制系统互锁跳闸。因此，选择性控制可以使系统保持连续运行，尽管可能处于次优水平。当一个过程可能超出某个限制或约束时，一个控制器可以从另一个控制器那里接管对被控变量的控制。

图 9-5 所示为一个实例，水泵由变速电机驱动以从井中抽水，并为客户提供恒定的水压。

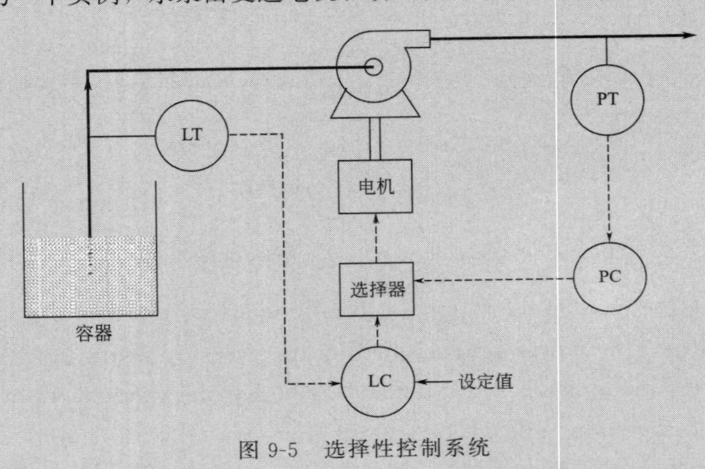

图 9-5　选择性控制系统

在这个系统中，一般情况下根据用户的用水量来调节水泵的转速，也就是以压力控制器来控制电机。

但是如果用户用水量的需求很高时，井里的水位会很低，甚至会导致水泵抽不到水。如果水泵在没有水通过的情况下运行太久，密封条就会损坏，这种情况是不允许的。解决这个问题的一个办法是安装一个液位开关来检测水位，如果水位太低，就关闭驱动水泵的电机。但这只是一个粗略的解决方案，因为它在保护水泵不受损害的同时，却要以完全切断用户用水为代价。

一个更好的解决方案是，当水位接近低水位时，让水泵减速以减少抽水量，这样可以保护水泵不会因抽不到水而损坏，同时也保持了对客户的服务。这被称为选择性控制系统。

通过将液位变送器连接到液位控制器，并使用选择器或功能块在压力和液位控制器之间选择较低的值来创建这样的选择性策略。液位控制器的设定值将设置在水泵连续运行的某个最低水位限度上。任何时候，只要水位低于这个设定值，液位控制器的信号将优先通过选择器，命令水泵减速，即使压力控制器要求更高的速度。液位控制器将取代压力控制器，优先考虑泵的使用寿命，而不是用户的用水需求。

Words and Expressions 词汇和短语

1. cascade control 串级控制
2. primary controller 主控制器
3. secondary controller 二级（副）控制器
4. phase lag 相位滞后
5. ratio control 比值控制
6. ingredient /ɪn'griːdiənt/ n. 原料
7. feedforward /'fiːd'fɔːwəːd/ n. 前馈
8. feedforward-feedback control system 前馈-反馈控制系统
9. selective control 选择性控制
10. override control 超驰控制
11. suboptimal /sʌb'ɔptiməl/ adj. 次最优的

References
参考文献

[1] 王树青，韩建国. 工业自动化专业英语 [M]. 北京：化学工业出版社，2001.

[2] Alan S. Morris. Measurement and Instrumentation Principles. Butterworth-Heinemann. 2001.

[3] Gregory McMillan.，& P. Hunter Vegas. Process/Industrial Instruments and Controls Handbook，Sixth Edition. McGraw-Hill Education. 2019.

[4] Douglase O. J. desa. Applied Technology and Instrumentation for Process Control. Taylor & Francis. 2004.

[5] Wolfgang Altmann. Practical Process Control for Engineers and Technicians，Newnes. 2005.

[6] Bela G. Liptak. Instrument Engineers' Handbook，Volume One：Process Measurement and Analysis，4th Edition. CRC Press. 2003.

[7] Bela G. Liptak. Instrument Engineers' Handbook，Volume Two：Process Control and Optimization，4th Edition. CRC Press. 2003.

[8] N. E. Battikha. The Condensed Handbook of Measurement and Control，4th Edition. International Society of Automation. 2017.

[9] Thomas A. Hughes. Measurement and Control Basics：Fifth Edition. International Society of Automation. 2015.

[10] William Dunn. Fundamentals of Industrial Instrumentation and Process Control. McGraw-Hill Education. 2005.